Principles of Innovative Design Thinking

Wenjuan Li · Zhenghe Song · C. Steve Suh

Principles of Innovative Design Thinking

Synergy of Extenics with Axiomatic Design Theory

Higher Education Press

Springer

Wenjuan Li
School of Transportation Science
and Engineering
Beihang University
Beijing, China

Zhenghe Song
College of Engineering
China Agricultural University
Beijing, China

C. Steve Suh
Mechanical Engineering Department
Texas A&M University
College Station, TX, USA

ISBN 978-981-19-0484-4 ISBN 978-981-19-0485-1 (eBook)
https://doi.org/10.1007/978-981-19-0485-1

Jointly published with Higher Education Press, China
The print edition is not for sale in China Mainland. Customers from China Mainland please order the print book from: Higher Education Press

This Springer imprint is published by the registered company Springer Nature Singapore Pte Ltd.
The registered company address is: 152 Beach Road, #21-01/04 Gateway East, Singapore 189721, Singapore

Preface

The book presents a comprehensive treatment on a novel design theory that fosters innovative thinking and creativity essential for addressing wicked problems. Wicked problems are ill-defined, ambiguous in both their aims and solutions, and complex with interconnected and intertwined (coupled) factors. While being ubiquitous and difficult, however, wicked problems share characteristics common to science and design in three regards; namely, agent finitude, system complexity, and problem normativity. These fundamental attributes allow a core cognitive process common to design and science to be identified and a strategic problem-solving conception of methodology to be formulated as a result. The novel theory presented in the book facilitates new opportunities for synergetic cross-disciplinary research and practice by incorporating the essences of the theory of Extenics to Axiomatic Design. Innovative thinking is enabled by exploring Extenics for problem reframing, paradigm shift, and abductive reasoning and by engaging Axiomatic Design in the co-evolution (iteration) of the need and viable design concept.

The theory is unique in that it is a framework for quantifying imprecise and vague design information available during the conceptual design stage as mathematical expression and algorithm early in the design effort and enables the objective evaluation and emergence of an optimal design concept from the multitude of viable ones. The principles of the theory help one articulate fundamental questions essential to tackling difficult design problems with regards to innovative design concept generation, proper thinking procedure, and guidance that drive design iteration. The principles are manifestations of Chinese philosophical thinking of abstraction using mathematical expression, functional mapping, and physical parameters. Through encouraging creative imagination, fostering broad abstraction, and promoting logical and structured inference, innovative design solutions are warranted.

A comprehensive investigation into the Axiomatic Design (AD) and Extenics allows the basis of the synergy of the two theories to be found that renders the mathematical description of the novel theory that is elaborated in the book. Functional requirement, design parameter, and design coupling as defined in the AD theory are described using the affair-element model, matter-element model, and relation-element model in Extenics. The issue of lacking unified and precise expressions in AD for elements in the problem domain and coupling description is resolved. Moreover, basic-element models for the process that maps the functional domain to the physical domain in AD are established. The search for initial conceptual design is transformed into one that addresses the feasibility of the design problem. Mapping from the functional domain to the physical domain is also made explicit. The needs for evaluating the independence of the elements in the design matrix, mapping and decomposing between domains, and expanding design solution space are also addressed. A general framework viable for proper design decoupling is also developed that allows viable decoupling methods to be formulated with ease in the context of the AD theory.

The book is conceived for students and real-world practitioners in engineering, natural and social sciences, business, and fine arts who seek to develop powerful design thinking for solving problems in a creative and innovative way. The principles presented in the present volume are universally applicable to problem-solving. They are not limited to specific disciplines or levels of studies. While the many illustrative examples included in the core chapters may seem suggesting that the book is designed for project-based engineering design courses, it is recommended that the methodologies be incorporated into teaching design thinking early in the curriculum, be it engineering or non-engineering.

The design principles of the book can be adopted as materials for a comprehensive coverage of conceptual design or for developing broad design thinking skill and capability effective for addressing tough problems that are wicked, difficult, and of a high degree of complexity. The book is arranged as follows. Chapter 1 introduces the essences of design thinking and contemporary engineering design methodologies. Chapter 2 defines the fundamentals including the definition for function structure, performance, constraint, and design requirements. Chapter 3 reviews the theory of Axiomatic Design and Chap. 4 gives a comprehensive account of the Extenics theory. In addition to investigating the complementary properties of the two design theories, the features and synthesis of the two powerful design theories are performed in Chap. 5. Drawing upon the creative synergy, Chap. 6 presents the innovative design methodology with the subsequent chapters, Chap. 7 documents the novel design decoupling methodology and Chap. 8 projects in brevity the bright outlook that the theory promises.

Beijing, China Wenjuan Li
Beijing, China Zhenghe Song
College Station, USA C. Steve Suh

Contents

Chapter 1
Introduction

Design is a process of creating a solution or a set of solutions viable for addressing a need through exploring science and physical principles. Knowledge from science, mathematics, and engineering practice along with creativity is employed to transform resources to meet the design objective with efficacy and measurable performance. The operation of designing artifacts and the construction of the designed artifacts have long been considered inseparable. However, designing of products and their fabrication in industrialized societies are separately pursued to better address increasing complexity and the cost associated with it. The process of building a design is initiated when the process of designing it is complete. Significant amounts of efforts have been devoted to understanding the design process and the ability to design in an innovative manner. This ability to apply knowledge to generate novel design solutions that satisfy specific goals and requirements has broad-ranging implications for the well-being of the entire human race.

Of the many rationales for needing to understand the design process, the first is driven by the increasing complexity of modern designs. The availability of novel materials and new technologies, the rapidly increasing knowledge base available for conceiving innovative solutions, and the ever-growing demand for shorter product turnaround time all present a daunting challenge to designers. Responding to these demands calls for a systematic approach that help enable creative design thinking and foster innovation as a fundamental attribute. In addition, the complexity of products and the short bringing a product to market time mandate that a team of designers work on multiple design tasks. A well-defined, structured approach to design is essential to organizing and coordinating team efforts in attending to satisfying all the requirements and avoiding making costly mistake and delay.

These needs have resulted in a plethora of design methodologies, procedures, techniques, aids, and virtual tools for dealing with activities that designers may use at different stages of the overall design process. They include rational procedures that formalize time-proven design practices and adaptations of methods from unrelated domains such as operations research and decision theory. All the design methodologies have the following two features in common–externalizing design

W. Li et al., *Principles of Innovative Design Thinking*,
https://doi.org/10.1007/978-981-19-0485-1_1

thinking and formalizing design processes into structured procedures. The former feature involves extracting the thought and thinking process out of the mind of the designer and organizing them using charts, tables, and diagrams. Externalizing design thinking is arguably very powerful when dealing with wicked problems and problem-solving requiring multiple of individuals working in teams, freeing the mind of the designer to pursue creative thinking. (See Chap. 4 for a discussion about problems with wicked properties.) The latter feature aids to avoid oversights when developing a design. Formalization also helps widen the search for appropriate design solutions. Research over the last several decades shows that engineering design process practiced by experts can be formalized in ways that can be easily learned by people who are inexperienced and less skilled. Research that increases our understanding of design process also makes the process explicit and accessible to anyone who is interested in learning it.

In the design process, design thinking is the thought process all great innovators practice for generating and implementing the basic notions that characterize a solution as being viable and innovative. It is a process for approaching design solution and innovation that center on addressing the true needs and the users for whom the solution is developed. The process can be explored to foster creative problem-solving in a systematic and structured way. It is particularly powerful in addressing difficult problems that are ill-defined and difficult to be solved using standard methods and approaches by re-framing the problems and through questioning assumptions and implications. The process can have a significant impact on the novelty, quality, implementation, value, and cost of the solution. The chapter provides an overview of the framework for the design thinking as well as engineering design process. It concludes with a brief section on the arrangement of the book.

1.1 Design Thinking

Design thinking as a burgeoning while yet to be fully defined concept has been seeing increasing interests in addressing wicked problems in areas beyond engineering design. Design thinking was first conceived to educate engineers to think and practice like a designer does. John E. Arnold, an engineering professor at Stanford, was among the few who first wrote about design thinking. In the book entitled "*Creative Engineering*," Arnold established that design thinking is creative problem-solving that differentiates itself from decision-making by meeting four requirements. It must be a better combination that is much more than just being something different. It must be tangible as in something that can be seen, felt or reacted. It must be forward-looking in addressing human needs. It must be of a synergetic quality; that is, having a multiplicative effect in achieving a value that is much greater than the sum of all the individual components combined. Design thinking has since evolved as a way of thinking gaining popularity not just in architecture and engineering but also science, business and art. Thinking like a designer is a way of viewing and responding to the world that materializes into various forms. In addition to creating artifacts and meaning, design thinking is also considered as either a reflexive practice, a problem-solving activity, a way of reasoning and sense making or a way of working. To business professionals, design thinking is a management theory that entails effective approach and skill for managers.

Design thinking is an iterative process by which one identifies the real need, challenges assumptions, and redefines problems so that viable solutions and solution strategies not immediately apparent with one's initial level of understanding can be formulated. The strategic, intellectual reasoning by which design solutions are developed is core to design thinking. Design thinking is a solution-based approach to problem-solving. It is the opposite of problem-based thinking which is characterized by focusing on constraints and restrictions. Need identification, problem framing-reframing, solution ideation, and creative thinking are among the broad set of subjects that design thinking addresses. Design thinking, as is presented in the book, imparts the abilities to resolve ill-defined or wicked problems, conceive solution-focused strategies, invoke abductive reasoning, apply physical principles, and establish quantifiable measures.

Design thinking enables designers to focus on developing an in-depth understanding of people's needs and desires for which design solutions are created, setting the path to better products, services, and processes. In engaging design thinking, user's need and desire are addressed with solutions that are both technologically feasible and economically viable. It also provides creative tools to ordinary people who are not professionally trained to be able to undertake challenging complex problems with confidence and with ease. The process starts with asking the proper questions that leads to the comprehensive discerning of the problem and the identification of the true need. To foster the generation of innovative solutions, the ideology that underlines design thinking encourages shifting of mindset and addressing problems from a new perspective. In addition to better understanding the needs of the users, design thinking also helps minimize the risk associated with launching new products and create outside-the-box, revolutionary solutions that emerge from shorter design iteration cycles.

1.2 Design Thinking Process

John E. Arnold Design presented a series of steps that bring this ideology that is design thinking to life–starting with building empathy for the users, right through to coming up with ideas and turning them into prototypes. The power of design thinking is in tackling wicked problems–not ordinary, common problems that have tried-and-tested solutions but rather complex difficult problems that seemingly have no solution at all. Not only are these problems difficult to define, but any attempt to solve them is likely to give way to even more problems. Wicked problems are ubiquitous, ranging from climate change and poverty to challenges in nuclear proliferation and geopolitical conflict to achieving sustainable growth in global business, just a few examples.

Fostering creativity, innovation, and paradigm-breaking and also highly actionable, design thinking is effective in resolving the wicked problems.

There are five phases that are pivotal to design thinking. Rather than a detailed prescription to follow, design thinking offers a set of guidelines and high-level principles that one can interpret as needed for addressing the true need.

Inspiration–The first step taken in the design process is understanding the problem essential for establishing objectives, benchmarks, requirements, technology needs and how the design solution or product will fit into the market.

Empathy–Empathizing with the users is the most important principles of design thinking. When creating design solutions, be them products, services or hardware, truly understanding the perspective of the user is imperative. Finding solutions that respond to user feedback is essential because people are the drivers of innovation.

Ideation–Design thinking is a solution-based framework focusing on coming up with as many ideas and equally viable potential solutions as possible using both divergent and convergent thinking. Ideation is both a core design thinking principle and a step in the design thinking process.

Experimentation and Implementation–Design thinking is an iterative approach requiring the repletion of certain steps in the process to address flaws and shortcomings in the early versions of the solution. Once a few of the best ideas are generated, modeling and digital or physical prototyping are created that can be tested, evaluated, and refined.

Testing–Test a digital or physical prototype on the users and see how they engage with the prototype is important for planning the next move forward. From these tests, the users' need and want are better understood. This is a phase for engaging the users in the design thinking process to co-create the solution.

1.3 Engineering Design Methodology

Creation of artifacts is one of the core functions of engineering with design being defined as the central activity. It is design that distinguishes engineering from the pure sciences. Engineering design is well summarized in a definition by ABET as follows: "Design is the process of devising a system, component, or process to meet desired needs. It is a decision-making process (often iterative), in which the basic sciences, mathematics, and engineering sciences are applied to convert resources optimally to meet a stated objective. Among the fundamental elements of the design process are: the establishment of objectives and criteria, synthesis, analysis, construction, testing and evaluation. The design component of a curriculum must include some of the following features: the establishment of objectives and criteria, synthesis, analysis, construction, testing and evaluation, development of student creativity, use of open-ended problems, development and use of design methodology, formation of design problem statements and specifications, consideration of alternative solutions, feasibility considerations, and detailed system descriptions. Furthermore, it is essential to include a variety of realistic constraints such us economic factors, safety, reliability, aesthetics, ethics, and social impact."

Engineers have a design mindset that is not just problem-focused, it is also solution focused and action-oriented requiring both analytical skills and imagination. They are trained to generate and provide engineering solutions as designs by engaging in engineering methods that follow the general steps outlined below which is an iterative process typically cycling through build, prototype, test, and redesign.

Idea–This phase always begins with a problem. Defining the problem is the most critical part of the phase, validate its value, and identify the user who desires its solution. The problem statement is typically vague with colloquial needs that do not indicate technical solutions. Execution of the phase is to perform an analysis to determine the viability and feasibility of the stated need. Viability suggests that there is significant value in pursing the solution. Feasibility serves as a check on whether the idea can be realized.

Concept–Concepts described through the use of mathematical model, physical principle, numerical simulation, simple drawings or sketches are generated in the phase, all of which are required to be viable and demonstrating that the solution meets the expectations and requirements put forth by the users. The viable concepts are developed by following established techniques and methodologies such as the Axiomatic Design, design parameter analysis, and TRIZ to evaluate all the salient features and recombine elements of one concept with elements from others in an effort to generate a single optimal concept. Merging concepts requires exercising design judgment and compromise, allowing viable concepts to emerge.

Planning–The planning phase is to define the implementation plan including identifying needed resources, tasks, task durations, task dependencies, task interconnections, and required budget. A wide range of tools are available for practicing the phase and conveying this information to relevant stakeholders including Gantt and Pert charts, resource loading spreadsheets, sketches, drawings, proof-of-concept models to validate that the design can be successfully completed. System engineering diagrams are often used to plan for the development of system solutions having many an interconnection of smaller and less complicated sub-systems. System engineering diagrams make explicit all the inputs and outputs for each individual component, module, and sub-system, as well as the way in which they transform the inputs into outputs.

Design–This phase is when the viable concepts are put to the test to see if they actually work. Details are specified and specifications are established. User requirements, inherent constraints and systems engineering models are translated in this design planning phase into engineering specifications that an engineer can work with to design and build a physical configuration for prototyping and testing.

Development–Engineering documentations such as schematics, drawings, source code, specification sheet, and other relevant design information is generated in this phase. Development is essential for supporting the realization of the working prototype, be it digital or physical, that demonstrates the viability and feasibility of the solution to the problem. The solution may be a tangible working prototype or an intangible working simulation. As it is commonly the case that nothing works the first time, so this part of the process is more iterative than the other phases. Specifically, it consists of the iterative cycle: design, prototype, test, and redesign.

While testing, prototyping, and redesign are considered separate phases, they are not independent of each other. They more often than not occur side-by-side with development as a design morph from a concept to an artifact with specific physical features embodying the viability and feasibility. Whether the concepts meet both the anticipated design specifications and the user's requirements of the solution is realized through testing. That is, the verification and validation of the design solution is achieved through experiments–an information-gathering method where dissimilarity and difference are assessed with respect to the design's present and compared to desired state for the design. If test results are in acceptable agreement or otherwise in conflict with the a priori stated behaviors are determined through experiment. When the test observations and results are in disagreement, it is necessary to identify the root causes and begin corrective action to resolve the discrepancies. Enough rounds of successful testing verifications and validations are essential to generate acceptable results and to reduce any risk that the desired behavior is presented.

Launch–In this final phase all testing is complete and all the working prototypes have demonstrated functionality. The engineering design and documentation package are then released to manufacturing facilities for production.

1.4 Engineering Design Process

Technological innovation, product design, product development, product realization, invention, and engineering design are terms commonly used to refer to the process by which products are created, developed, and delivered to the users. While practicing professionals and academics often differ on the definitions of these terms and also on the many different models of engineering design process, there are common elements that are shared and agreed upon:

- A stage of identifying and analyzing the true need prior to initiating conceptual design
- A stage of conceptual design for creating new ideas that satisfy the need
- Activities through which a concept is turned into an overall product or system layout
- A stage of finalizing the design details

A high-level view of the engineering design process is represented by the chart shown in Fig. 1.1. Considerable attention and emphasis are placed on Design Methodologies and the strategy associated with each of the four stages and on the philosophical basis for the methodologies that are to be employed in achieving the desired goal of each stage. The activities in each design stage define the central aspect of the stage. The overarching purpose is to enhance the efficacy of executing the activities within each stage while fostering the potential for innovation.

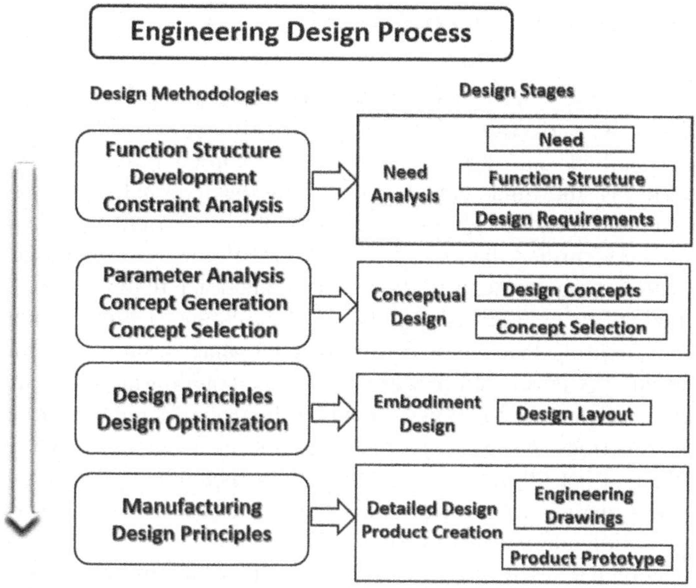

Fig. 1.1 Engineering design process

The design process can be organized into four major stages, namely, need analysis, conceptual design, embodiment design, and detail design. This organization of the design process follows from Pahl and Beitz (1996). The right column in Fig. 1.1 tabulates the four stages and the outputs from each of these stages. In the first stage design methodologies including Function Structure Development and Constraint Analysis are employed to perform Need Analysis that generates Need Statement, Function Structure, and Design Requirements as outputs. In the same token, Parameter Analysis, Concept Generation and Concept Selection are followed in the second stage to generate and select concepts. The methodologies in each stages provide the framework within which design activities are performed and the qualities of the resulted outputs are evaluated. The design methodologies associated with each stage are a result of considering different sources of brilliant contributions. The work by Pahl and Beitz (1996) provides an approach to function structure development, the concept-configuration model can be inspired by Jansson (1990), and the concept selection process can be adopted from Pugh (1996). Being the primary focuses of the book, the Need Analysis and Concept Design stages are discussed in detail in the book.

1.4.1 Stage of Need Analysis

Need Analysis is the stage where the true need is extracted from a colloquially expressed need. By performing abstraction and critical parameter identification, the problem is "distilled" into a need statement that captures, in technically precise yet abstract terms, the fundamental issues in the design challenge. (See Chap. 2 for a discussion about need identification and analysis.) Various functional requirements (FRs) are identified that must be met to satisfy the need. In addition to FRs, inherent constraint requirements (CRs) and non-functional requirements (NFRs) are also identified. NFRs are the size, weight, cost, regulations, environmental mandate or operation condition that the design must not violate. Using the information, a function structure is developed that establishes the solution-independent functions and sub-functions that are to be performed by the design. Function structure tabulates all the functions that a design must perform to achieve the need. Based on FRs and NFRs, a list of design requirements is generated. A properly performed need analysis generates a thorough understanding of the design requirements and the outputs that dictate the rest of the design process. Care must be taken at this stage to maintain the independence of FRs, as coupled FRs invariably result in poor designs and significantly increase product development time and cost (Suh, 1990). Also, the solution space needs to be kept as wide open as possible to not preclude any solutions through misperception, unwarranted assumption, or false constraint. Need analysis is important in that it lays the foundation for the design and foster innovative thinking.

1.4.2 Stage of Conceptual Design

The second stage in the design process is conceptual design. The designer will always need be inspired to be able to come up with innovative initial ideas. These ideas can come from the step we call technology identification or from recognizing critical issues during the need analysis. Conceptual design involves searching for the scientific principles and technologies that can potentially be exploited to satisfy the design need. Conceptual solutions are then derived from applying the principle and technology that are selected. In the conceptual design stage, the designer typically develops more than one solution to increase the likelihood of innovation. These solutions are then evaluated against each other to determine which one will be developed further. This stage as depicted in Fig. 1.2 is known as concept selection.

Note that the model recognizes that the interfaces between conceptual design and realization, and between need identification and analysis and conceptual design, are bidirectional. That is, it is possible that a conceptual design will be "sent back to the drawing board" or rejected altogether if the results of the downstream efforts reveal

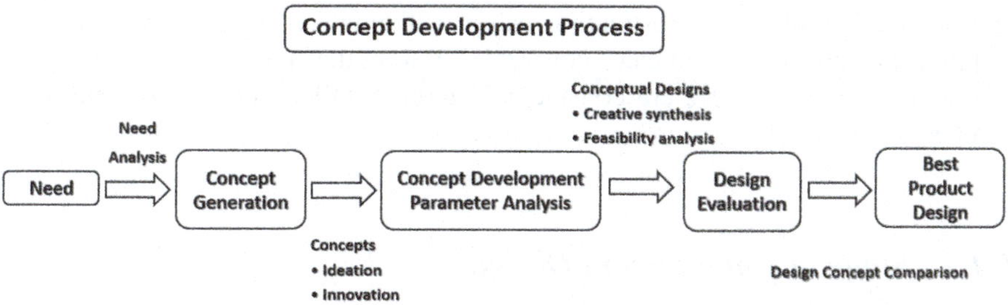

Fig. 1.2 Concept development and selection

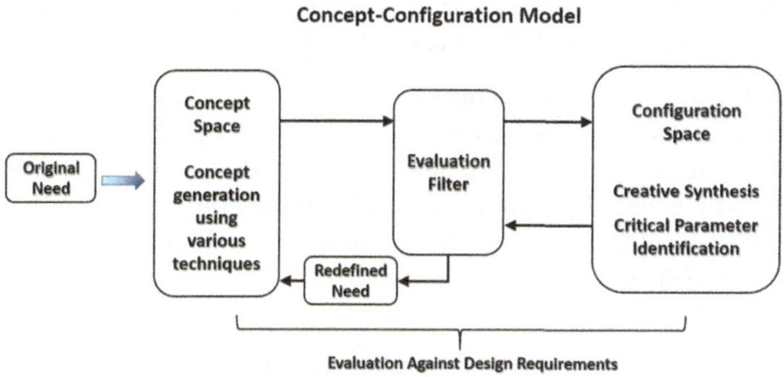

Fig. 1.3 Concept-configuration iteration model

issues overlooked during the need analysis and conceptual design stages. In other words, a bad conceptual design cannot become a good design regardless of the quality of the realization effort and may therefore call for additional work. Lengthy delays and increased product development costs can be effectively mitigated by following the iterative process shown in Fig. 1.3 to avoid any subsequent adverse effect that could impact the success of the design solution.

Each generated concept is developed into a configuration and is evaluated for satisfying the requirements of the design. The concepts are either developed into viable solutions or discarded by using the concept-configuration model. Conceptually different solutions are generated and each iterated through the concept-configuration looping process till an objective evaluation of the concept is achieved. The output of the conceptual design stage is a viable conceptual design that is selected based on a thorough evaluation using both the functional and non-functional requirements of the design.

Because the high cost associated with amending conceptual flaws in the embodiment and detailed design stages, success of the design process hinges on the outcome of this stage. One of the strengths of the Principle of Design Thinking is in providing a systematic methodology to develop an initial, ill-defined rough idea into many a viable conceptual solution. Because the principles are manifestations of Chinese philosophical thinking of abstraction using mathematical expression, functional mapping, and physical parameters, conceptual designs developed through exercising and applying the principle are considerably more detailed and quantifiable using numbers in this book.

1.4.3 Stage of Embodiment Design

In this stage, form and shape are given to the selected conceptual design. Concepts are developed into an assembly or layout that defines the relative positions of various components, their sizes, shapes and interrelationships. The following design principles are applied when giving shape to the concept:

- Separate function (Avoid coupling)
- Provide direct and short transmission path
- Constrain to required degree (Do not over-constrain)
- Minimize gradients /Match impedances (Let Form Follow Function)
- Provide functional symmetry / Balance forces and moments internally
- Design for self-help
- Design to fail-safe

During this stage, design layout is optimized by balancing the demands for functionality, safety, and manufacturability. If the design is having coupled FRs, this is the stage where the conflict between functional requirements is to be addressed and resolved to prevent compromises in the performance of the final product. The above principles can also be used to evaluate an existing design and suggest improvements to a design.

1.4.4 Stage of Detailed Design

In this stage, drawings are prepared from the layout or assembly drawings, and material and manufacturing processes are specified. In this stage, the designer performs detailed analyses of the design via computer simulation and/or prototype testing. The results assess the performance of the design and evaluate how well it meets the original requirements. The designer, in conjunction with the manufacturing facility and utilizing manufacturing design principles, develops detailed drawings of the design. At this stage, the design is in its final stages and fundamental improvements to the functionality of the design are very difficult and expensive to achieve.

1.5 Arrangement of the Book

This book uses examples to elucidate the principles and methodologies presented, and the examples are elaborated in ways that clarify the thought processes followed. Readers are encouraged to study the examples thoroughly to establish a comprehensive understanding of the methodology by analyzing and exploring each part of the thinking process in the step-by-step manner that is presented in the book. When the methodology becomes more familiar, the formalism of this approach will be less important. However, successful, high-quality concepts and viable design solutions come about through practice and hard work, as well as inspiration. It is often argued that clever, creative conceptual design skills are a gift, not something that can be acquired by learning. However, the authors have seen first-hand the power of the principles which synergize scientific thinking with high-level abstract (conceptual thinking). Just as an incredibly gifted athlete must practice to reach the highest level of competitiveness, so, too, can designers benefit from the methodology presented herein that sharpens their ability to think in ways that foster creative innovative design solutions. The authors have considerable experience in design thinking and engineering design methodology to college students and practicing engineers alike. This approach to conceptual design has been tested many industry sponsored design projects. The combination of theoretical discussion of the principles and the use of illustrative examples has proved to be an effective way to learn the novel methodology.

Chapter 2 introduces a methodology for identifying and analyzing real (true) needs as the preparatory step to the synergistic design theory. While Chap. 3 is an introduction to the Axiomatic Design theory, Chap. 4 is a detailed treatment on the Extenics theory employed with illustrative examples to illuminate the essence and quality of the methodology. Chapter 5 presents the basis for the creative synergy of the Axiomatic Design theory with Extenics with a focus on transforming narratives into semantics that inspire a broader set of viable concepts. An expanded discussion is also provided along with specific guidelines for implementing the methodology. Chapter 6 introduces the enhanced design principle that explores the essences of the Axiomatic Design theory (AD) and Extenics. As discussed in Chap. 5 the two theories are mutually complementary. The vague description of elements and ambiguous domain mapping are improved by incorporating Extenics into AD making use of mathematical expressions and semantic of logic. Starting with identifying the lack of a uniform description for design coupling and the need for a systematic methodology for decoupling, Chap. 7 explores a novel method that allows problems involving design coupling to be described by their attributes using the basic concepts of Extenics. The resulted descriptions are well-defined in the basic-element form with a set of uniform characteristics, enabling the decoupling process to be general, structured and effective in search for the knowledge needed for solving the coupling problem. Examples are followed to illustrate the validity of the decoupling method which is a synergy of Extenics with AD. The book concludes with Chap. 8, which provides an outlook on how to proceed with applying the design principles and process to address a broader set of wicked problems. Some of the cognitive aspects of the principles are also discussed.

References

Jansson, D. G. (1990). Conceptual engineering design. In M. Oakley (Ed.), *Design management: a handbook of issues and methods* (pp. 219–230). Basil Blackwell.

Pahl, G. and Beitz, W. (1996). *Engineering design: a systematic approach.* Springer.

Pugh, S. (1996). *Creating innovative products using total design.* Addison Wesley Publishing.

Suh, N. P. (1990). *The principles of design.* Oxford University Press.

Chapter 2
Need Identification and Analysis

In engineering design, identifying the real need is generally the first step. The particular step is essential to not just generating viable solutions to the need, but also providing form and shape to the solutions and enabling the conclusion of the entire design endeavor by building and testing and validating prototypes. The process involves organizing and managing human as well as physical resources. Relevant issues such as cost, safety, reliability, ethics, and social impact are also considered during the design process.

Need identification and analysis together set the stage for conceptual design. The effectiveness of the conceptual design process depends on how well a need in the marketplace is understood. The two major functions of need identification and analysis can be described in broad terms as: (1) Discovering what the real need is and (2) Analyzing the need such that the best possible solutions to meet its requirements are not precluded by the way the need is understood or described. The primary roles of the need identification and analysis are to eliminate bias from the process. Identifying the real (true) need deals with recognizing and removing preconceptions of the need definition. Furthermore, ensuring that the best possible solutions are not inadvertently ruled out of the process is accomplished by analyzing the need in a way in which the influence of preconceived solutions to the problem is objectively removed from the analysis. These two influences on the whole process are inherently quite similar and suppress the natural tendencies of human problem solvers. Therefore, the overt process of ensuring objectivity at this early stage is an essential part of good engineering design practice.

2.1 Need Identification

When presented a problem, skilled designers explore human cognition and reasoning to extract the core issues while inexperienced designers overwhelmingly appeal to known solutions and prototypes for inspirations and ideas. Engaging cognitive

W. Li et al., *Principles of Innovative Design Thinking*,
https://doi.org/10.1007/978-981-19-0485-1_2

thinking to identify the real need is a trait common of all experienced designers, a powerful skill enabling the generation of novel solutions. The skill can be developed by consciously engaging Abstraction to think on an abstract level, applying Critical Parameter Identification to considering critical down-stream issues early, and practicing Questioning to escape fixation.

2.1.1 Abstraction

Abstraction is the process by which a perceived need is progressively transformed from a colloquially expressed statement into a functionally precise definition, using technically fundamental terms. Abstraction enables the underlying issues of the need to be extracted, thus allowing the gaining of insight into the tasks that must be performed. Being able to think abstractly is considered one of the essential skills required of all designers.

Consider the task of designing a transportation canal to facilitate the movement of ships across a landmass, with "Connect two bodies of water with no gradient" been given as an initial need statement. The Suez Canal that connects the Red Sea to the Mediterranean Sea at sea level is an example of solution satisfies the need statement. However, "Is no gradient a real constraint?" Invoking the abstraction process brings one to the realization that many canals with gradients exist and enclosures with gates are used to raise and lower ships during their passage through the canal from level to level. For example, the Panama Canal uses locks that raise and lower the waterway in the canal by 85 ft during a single passage. It is evident that "no gradient" is a false constraint.

The abstraction leads to the emergence of the revised need statement: "Connect two bodies of water." Do two bodies of water really need to be connected with water? Obviously, the answer is negative as the real need is to move ships across a landmass. The revised need broadens the solution domain to include the idea implemented in 1968 by the Ronquières Inclined Plane, a Belgian canal inclined plane on the Brussels-Charleroi Canal, where boats are transported in water-filled railway containers across a 1432 m stretch with an elevation of 67.73 m vertically. The example exemplifies the power of abstraction in expanding the solution space and in aiding the envisage of innovative solutions. The level of abstraction is reached by evolving the given problem statement through:

- Eliminating solution-specific detail
- Defining the problem in solution-neutral term
- Transforming quantitative information into qualitative information
- Questioning and invalidating false constraint
- Increasing the technical conciseness of the statement by defining the various terms used in the need statement and looking for scientific principles that are relevant to the particular solution

Need Statement

Colloquial

- To stop a car
- To reduce the speed of a car from 60 mph to 0 mph in less than 200 feet
- To decelerate a car at a controlled rate
- To dissipate the kinetic energy of a car at the required rate
- To transform the translational kinetic energy of a car at the highest acceptable rate

Abstract

Fig. 2.1 Abstraction of need statement for car brakes design

The validity of abstraction is evaluated by answering questions such as, Is the abstraction technically precise? Is the abstraction simultaneously more general yet less vague? Does the abstraction capture the real need? Is it solution independent? Are all the possible solutions included by the need statement? Are extreme cases also included in the solution space? In following the abstraction process innovative and non-traditional solutions are encouraged as the direct result.

Another example is designing the brakes for a car. Figure 2.1 shows the evolution of the need statement from initially being colloquial to eventually more technically precise but more abstract as the elimination of solution specific details makes way to the emergence of qualitative terms. The resulting final need statement is simultaneously technically precise, solution independent, general but not vague, implying a wide range of feasible solutions.

2.1.2 Critical Parameter Identification

Critical Parameter Identification is the systematic process by which a designer identifies the crucial "make-or-break" issues in an identified need. These issues can be physical, natural, chemical, or mathematical concepts that are relevant to the need. Critical Parameter Identification is closely associated with Abstraction. One of the prominent characteristics shared by skilled designers is the ability to reach the core issue of a problem by identifying the underlying critical parameter (Cross, 2000).

Consider the example of the car braked design. Figure 2.2 shows the evolution of the need and the associated critical parameter at each abstraction stages. The initial need is to reduce the speed of the car from 100 to 0 km per hour in less than 60 m with "stopping distance" being the critical parameter. As the need is expressed in more abstract terms, the solution is to decelerate the car at a controlled rate, with "deceleration" of the car being the critical parameter in this stage. Upon further

Evolution of Need	Evolution of Critical Parameter
• To stop a car	
• To reduce the speed of a car from 100 km/h to 0 km/h in less than 60 m	Stopping distance
• To decelerate a car at a controlled rate	Deceleration
• To dissipate the kinetic energy of a car at the required rate	Energy dissipation rate
• To transform the translational kinetic energy of a car at the highest acceptable rate	Energy transformation rate

Fig. 2.2 Evolution of critical parameter for car brakes design

abstraction, and considering both disc and drum brake solutions, the need statement is, "To dissipate the kinetic energy of the car." Further questioning the need makes explicit that "dissipation" is an unjustified restriction. It is but one out of many viable ways for reducing the kinetic energy of the car. This realization results in a revised need that states "To transform the translational kinetic energy of the car at the highest acceptable rate", with the term "translational" identifying the kinetic energy that must be transformed. By identifying the critical parameter, the required information for meeting the need is identified. The "energy transformation rate" indicates that the total energy of the car is to be accounted for to meet an established rate limitation.

The actions taken to identify the critical parameter of the car brakes design include

- Identify the primary functions and primary constraints in the need where "transform the translational kinetic energy" is the primary function and "highest acceptable rate" is the primary constraint.
- Identify transforming kinetic energy and the transformation rate as the defining parameters for the primary function and the primary constraint.
- Develop (constitutive) relationships between variables that define the function(s).
- Identify the critical parameter through considering consequences of failing to perform the primary function.

Critical parameters generally come from two sources: limiting condition and gradient that defines the rate of change of a variable. Critical parameters are oftentimes determined by the limiting conditions at which the functional requirements of a design at interfaces with the environment. For instance, in the car brake example, the road conditions of the tire-road interface provide the limitations on the rate at which the energy transformation can be achieved. Nature responds to disruption by minimizing the induced gradients that jeopardize stability. Consequently, critical parameters often involve spatial or temporal gradients or slopes. "Energy transformation rate" in the car brakes example is a gradient-based critical parameter. As the magnitude and degree of difficulty of a design task is dictated by rate requirements, minimizing rate of change while meeting the specified need is central to the entire design endeavor.

2.1.3 Questioning

Abstraction is the process of considering the true need independently of their associations, attributes, or concrete accompaniments. Questioning the obvious is the determining factor to extracting the true inner meaning of a need. Skilled designers engage the act of abstracting and identifying critical issues (parameters) by asking questions. In the example on canal design, questioning the constraint of "no gradient" led to the rejection of the constraint as being false and unnecessarily restrictive. Similarly, in the example on car brakes, questioning the need concluded that dissipation of kinetic energy is exclusive in the handling of the kinetic energy. Realizing that energy can be transformed into other forms resulted in identifying transformation of energy as the true need. Storing and manipulating the transformed energy became options that would not otherwise have been considered. Because identifying the true need is a process of cognition for the acquisition, storage, manipulation, and retrieval of information, questioning is therefore an indispensable part of need analysis. Questioning need and assumption facilitates innovation by not getting fixated on certain ideas. Questions that aid in exploring the task and gaining insight into the true need should start with five "WH's" and "HOW", along with their respective opposites "WH not's" and "HOW not", as given in Fig. 2.3. What? When? Where?

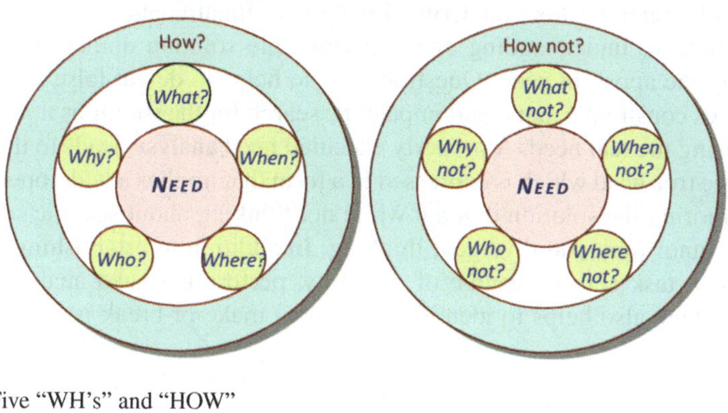

Fig. 2.3 Five "WH's" and "HOW"

Who? and Why? are arranged in that specific order to help with effective information gathering. Encompassing all other questions, How? is the means by which answers to WH's are implemented.

One of the primary functions of questioning is to point design thinking toward new domains to prevent fixation on existing solutions. The application of heuristic thinking in the thought process enables innovation and the examination of relevant issues yet to be considered. These heuristics are (a) considering contradictions (i.e., instead of making a system simpler, make it complex), (b) taking things to extremes (e.g., maximizing, minimizing, using multiple objects instead of a singular, trying radical arrangement), and (c) viewing adversity as an advantage (i.e., converting undesired problem into a desired solution or eliminate the undesired feature.) Questioning, along with the use of heuristics, is powerful in that it provokes thinking in new and different perspective.

2.2 Need Analysis

Before initiating the design process, the objective of the design challenge must be defined. Various design tasks are gathered from a variety of sources including customers and the marketing department. These sources usually provide their needs, likes and dislikes in non-technical and general terms. These perceived needs and preferences are reformulated into design objectives to allow the real needs to be extracted. Novice designers incline to conceive solutions right away based on their perceptions of these needs before gaining an understanding of the scope and magnitude of the design task. This urge to act is consequential in resulting in poorly defined design tasks and thus solving the wrong problem. Identifying the design tasks with comprehension is vital not just for clarifying the design needs, but also facilitating the efficient development of innovative and effective solutions.

Need analysis is a systematic process of identifying the real need. The design tasks that must be performed by the solution to address the need are also defined in the process. An in-depth understanding of the problem domain can be established by following Abstraction, Critical Parameter Identification, and Questioning. A comprehensive understanding of the permissible solution domain is gained by questioning the apparent need. Questioning also helps to defeat false requirements and fictitious constraints that could impact the search for the solutions that are viable for addressing the real need. A properly executed need analysis leads to the identification of the true need which is expressed in a form that implies a wide-open solution space. Exploring the solution domain while not thinking about specific solutions is the key to innovation and divergent thinking. In addition to establishing the scope of the design task and the degree of difficulty, performing need analysis early in the design stage also helps to identify the critical make-or-break issues, determine

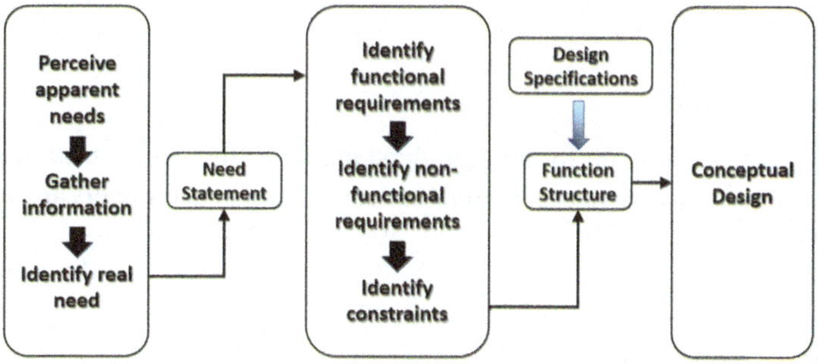

Fig. 2.4 Activities and outputs in the need analysis stage

the available information and required resources, establish the interfaces between design tasks, advise design team organization, and reveal the scope for innovation. Need analysis is the anchoring stage upon which the entire design process follows. The level of innovation and the breadth of the possible solution space for the overall design task is dictated largely by the skill in executing need analysis to identify the real need.

Figure 2.4 illustrates the activities in the need analysis stage. The outputs of the design activities include need statement, function structure, associated design parameters and constraints, and design requirements with the last two outputs used for evaluating the developed conceptual solutions in the conceptual design stage.

2.2.1 Perceive Apparent Needs

The first activity in Need Analysis is to determine the actual need. This is done by subjecting the perceived design need to Abstraction, Critical Parameter Identification and Questioning where the provided colloquial need is examined, reviewed, and validated. It is generally the case that the perceived need does not represent the true design challenge that must be addressed, thus not implying the solution domain.

2.2.2 Gather Information

Once it is determined that a valid need exists, information about the need is collected by following a coherent process of questioning using the five "WH's" and the "HOW" as follows to clarify the design task:

- What is the purpose of the design?
- Why is it needed?
- When is it needed?

- Where is it to be used?
- Who will use it?
- How will it be used?

In addition to gaining an in-depth understanding of the design task, the answers obtained to the above questions are vital to identify critical parameters and to perform abstraction.

2.2.3 *Identify Real Need*

The true need that defines the fundamental design task is determined in this stage. Abstraction, Critical Parameter Identification, and Questioning are followed to identify the cause and effect of the problem and to diagnose the task in a form that can be used to obtain the information that set both the magnitude and the scope of the design task. In so doing, the cause of the problem is addressed instead of solving the symptoms. Recall that transporting ships across the landmass was identified as the true need for the Ronquiéres Inclined Plane in Belgium. Constructing a canal is but a possible solution out of many.

Typically, the colloquial problem statement tends to be configurational, identifying the required end product rather than emphasizing the need, which is the function the solution must perform. "Design a robot arm," "Design a crane," "Design a heat exchanger" are some typical statements that calls up existing configurations that were the consequences of particular design solutions for a more fundamental need. They trap designers by subjecting them to fixating on similar solutions, thus inhibiting innovation. To defeat the trap, the perceived need must be abstracted to uncover the underlying real need–the one unwittingly expressed but not always recognized by the customer, who often adopts an analog to an existing solution to express the need. For example, a shaft is needed to transmit torque. However, transmitting power is a better need statement for the shaft which implies a broader set of potential solution including using an electrical wire to transmit power from a primary source such as an engine to a motor. By redefining the need, the solution set is effectively widened.

By performing abstraction, critical parameter identification, and questioning over the information gathered, the problem is refined and filtered into a *need statement*. This statement identifies in qualitative terms the primary design objective and the primary constraint must not violate. 'What must be done' by the design and not 'how it should be done' is implied in the need statement in which 'what the design must perform' is identified. An important characteristic of the need statement is that it is solution independent.

Need statement consists of a primary function and a primary constraint. The primary function can be identified by Abstraction and the primary constraint can be identified by Critical Parameter Identification. Consider the case where "Seal against leakage around a shaft without contact" is abstracted from "Design a labyrinth seal," an apparent need. "Seal against leakage around a shaft" is the primary function and the primary constraint is to do it "without contact." Another example is the car brakes problem where "To transform the translational kinetic energy of a car" is the primary function and "at the highest acceptable rate" is the primary constraint. The first part of the statement is concerned with "what needs to be done to what." It defines the primary functional requirement of the need, which is typically stated as a verb-noun (V–N) pair. The second part of the need statement identifies the critical parameter that defined the bound of the conceptual domain with which design solution are generated. Primary constraint can be perceived as the requirement that ultimately sets the magnitude of the design task. A fully developed need statement carries with it the knowledge that must be obtained before a conceptual solution should be embodied. Essential for defining the order-of-magnitude and scope of the design task, this knowledge thus also provides a qualitative characterization of the design task. In the car brakes example, the kinetic energy to be transformed and the acceptable deceleration rate are the two orders-of-magnitude that define the envelope of the solution space. Neither quantity is easy to obtain, however. The total amount of energy that needs to be transformed depends on the initial velocity and the mass of the car. The highest energy transformation rate must be given as a limit dictated by the physical ability of the driver and passenger to withstand the maximum deceleration. The limit is also impacted by an indeterminate condition that is road-tire interface. To maintain control at all times, an upper bound of the maximum and average deceleration of the vehicle must be set. There may be other factors that impact the bound of the highest acceptable rate, with some emerge subject to certain conditions and some under others. Because allocating resources and garnering required information needed can be demanding of one's time and effort, it is important to identify this information early in the design. Arranging for on-time delivery of the information is equally important to ensure that the design is not delayed by lack of knowledge.

There are occasions where it seems justified to have multitudes of functions and constraints be included in a need statement. However, the mandate that the need statement should feature only the primary function and the primary constraint obliges contemplating through the design challenge to identify the key issues and the scope of the design to ensure a comprehensive understanding of the total design task at hand.

2.2.4 Identification of Functional Requirements

Once the need statement has been defined, the next step is to identify the various functions that must be performed to satisfy the need. Just as a technical system can be divided into sub-systems and elements, similarly an overall function can be broken down into sub-functions of lower complexity as necessary to support the primary function. The main objectives of breaking the function into sub-functions are to

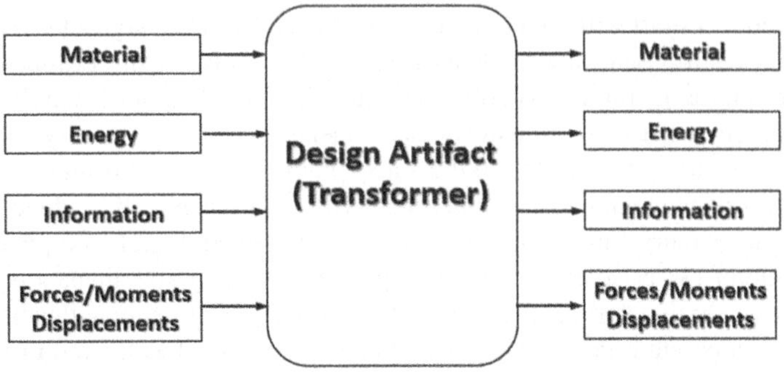

Fig. 2.5 The transformer model

- Simplify the design solution into smaller subtasks.
- Allow distribution of sub-tasks among various teams, and
- Facilitate the subsequent search for solutions.

A subfunction is a solution specific function or a sub-task to be performed in accomplishing a primary function. The subdivision continues until each subfunction is broken down to the final state with the primary constraint and design parameter. A constraint is a limiting factor that cannot be violated but is of paramount concern in the formulating of the design solution. A design parameter is determined to satisfy the corresponding functional requirement.

The decomposition process enables the emergence of innovative solutions to account for all the primary constraints and parameters in the overall solution.

One of the methods used in developing of sub-functions is the Black-Box Method (Cross, 2000; Pahl & Beitz, 1996), also known as the transformer model, as shown in Fig. 2.5. A design artifact can be considered as a transformer that takes in a set of inputs—energy, material, information, forces/moments and displacements and transforms them into a set of desired outputs. Information about the necessary inputs is gathered by asking questions on the inputs: What is the type of energy that is required? How much? What materials are required? What is the quantity of the materials? Is there any information that is to be taken in as input? Are there any forces or displacements that are to be considered? Similarly, information on the outputs is gathered. This sets the stage for using the Black-Box approach and assesses how the device transforms the inputs to the required set of outputs.

The typical breakdown of the functions and the organization of sub-functions into various levels is known as the function structure. Function structure is the hierarchical dependence of the need to the primary functions, primary functions to the subfunctions, the subfunctions to the parameters and constraints. The primary level or first level functions are those that must be performed to satisfy the need statement regardless of the solution that is finally adopted. Each of the first level sub-functions

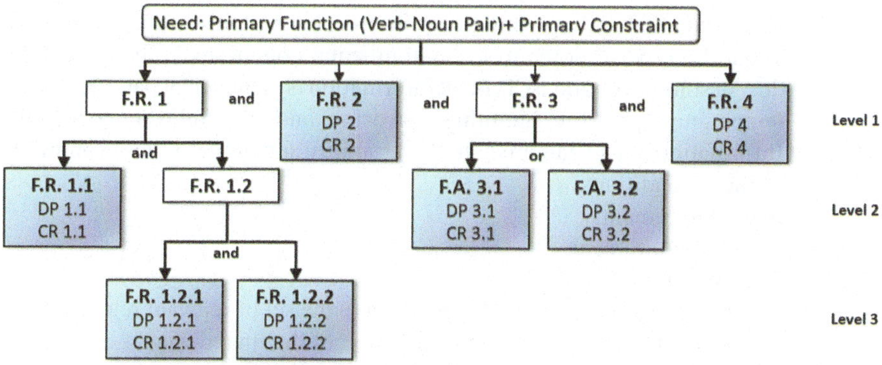

Fig. 2.6 A Generic function structure

is further broken down into second levels of solution independent functions, as shown in Fig. 2.6 where a branch graph is utilized to show how the problem is broken down. The first level contains the need statement, and the next two levels feature the primary functions and the supporting subfunctions, respectively. Functional alternatives (#3.1 and #3.2) are functions that satisfy the previous function (#3) by performing either one of the alternatives. These functional alternatives arise only when there are a small number of choices and each of these choices is at a level where they are still solution neutral. The lowest level functions are those for which further decomposition results in, at least, some solution dependent sub-functions. The goal is to build a solution independent function structure to the lowest level possible. It is seldom necessary to proceed beyond the third level. Below each of the subfunctions, the design parameter (DP) and constraint (C) for that particular subfunction are identified. The constraints and the design parameters for the functions are explained in the following sections.

2.2.5 *Identification of Non-Functional Requirements*

Non-functional requirements (NFRs) are solution-independent requirements must not be violated while meeting functional requirements. Non-functional requirements are frequently overstated. They are constraints imposed by external considerations that are difficult to challenge. Because they are negotiable, NFRs must be thoroughly questioned before categorizing and ranking them. The followings are sample categories that help address design issues and gain insight into the problem.

1. Non-functional performance requirements–The operation boundary of the design task is identified using sub-categories such as

   a. Environmental and operation conditions–This pertains to the information about the type of domain or environmental conditions within which the design solution operates. Typical information is on physical parameters like temperature, pressure, humidity, corrosion, and other relevant conditions. Information is also gathered on the variability and range of these conditions of these conditions.
   b. Size requirement.
   c. Weight requirement.
   d. Time restriction.
   e. Regulatory requirements–This pertains to the set of relevant regulations to be followed for the testing and validation of the final design.

2. Safety requirements–Safety is of primary concern with design. Another category is the impact on environment such as emissions and effluence control.

   a. User safety requirements–These are best listed in hierarchy: must not kill, must not cause bodily injury, must not harm any person in any.
   b. Environment safety requirements–The list includes the following entries: must not harm the immediate environment, must not cause harm to environment that may affect the user, must not cause harm to operating environs within the system.

3. Cost–It is a common practice to be conscientiously related cost to value at the early stage of design. Cost requirements include:

   a. Cost of the design.
   b. Operating and maintenance cost.
   c. Disposal cost.
   d. Lifecycle cost.

4. Ergonomics.
5. Aesthetic appeal.
6. Manufacturability and assemblability issues.
7. Maintenance and operational issues.
8. Life cycle, durability, and reliability.

While not all the above categories are relevant to the design task at hand, identifying if there are any showstoppers, i.e., hard constraints, is recommended.

2.2.6 Identification of Constraints

Constraints are limits imposed on functional requirements that define the envelope within which the functions must be satisfied. They define the degree of difficulty of a design task. Constraints invariably restrict the solution space. As such separating the real constraints from the false ones is vital to fostering innovation. As noted earlier non-functional requirements oftentimes act as constraints on the operation and implementation of functions.

2.2.7 Organization of Function Structure

In this activity, the functions, together with the respective constraints and design parameters, are organized into various levels. The generic function structure shown in Fig. 2.6, where CRs were the constraint requirements and DPs were the design parameters for the respective functional requirements (FRs). Design parameters are technical characterizations of the function and/or constraint. These provide the units for the functional and constraint requirements. Order of magnitude calculations provide the magnitude of the design parameters.

Function structures help facilitate innovation and aid in identifying interfaces. A good function structure has two major qualities. The first is that the lowest level functions are solution independent and the second is that, as dictated by the first design axiom, independence of functions is maintained. Independence of functions requires a one-to-one mapping of design parameters and functions. The general characteristics of a function structure are that when one moves upward in a function structure, it reveals as to why a function is required, and when one moves downward it answers as to how a FR is met. All FR's in a function structure must be performed by all the solutions.

The function structure contains the various functional requirements (FRs) with the associated constraint requirements (CRs) and the design parameters (DPs). By repeated questioning on the FR's, CR's and DP's the designer gains insight into the design task. With additional information gained, the function structure evolves and refines.

2.2.8 Development of Design Requirements

Design requirements are the overall requirements that must be satisfied. They are the evaluation criteria against which conceptual designs are assessed. These are based on the performance requirements plus on the non-functional requirements. They identify as to what the design solution must do and how well it does it. The requirements should be quantified as precisely as possible to avoid ambiguity, misunderstanding and potential conflict later in the design process. Design requirements associated with the FR's are derived from the constraint requirements and design parameter assigned to each function.

2.3 Progressive Cavity Pump Example

When crude oil wells are drilled there is usually significant pressure in the reservoir being tapped. With this high pressure the well is often free flowing, meaning that the fluid is high enough to overcome the static head and the well will produce without the use of external aids. When the fluid head drops below the static head, the well will require artificial lifting to continue producing oil. Progressive cavity pumps are often used to lift crude oil. Invented by Rene Moineau in 1929, progressive cavity pumps are efficient and widely used due to their ability in handling a wide variety of fluids with a constant flow rate. Unlike other types of pumps, progressive cavity pumps can move virtually any type of fluid from a clean, low viscosity fluid like oil or water, to delicate products like whole cherries, to very abrasive or highly viscous fluids.

Progressive cavity pumps in their fundamental configuration consist of a single threaded helical screw, or rotor, rotating eccentrically inside a double threaded helical nut, or stator. As the rotor turns inside the stator, cavities are formed that progress from one end of the stator to the other end. In one revolution of the rotor, two separate cavities are formed, one cavity opens at the same rate as the second cavity closes. The positive displacement movement results in a predictable steady flow. In most pumps, the stator is formed with an elastomeric material that fits on the rotor with an interference fit. The compression fit between the rotor and stator results in the formation of seal lines where the rotor contacts the stator, thus assuring separation of the individual cavities progressing through the pump with each revolution of the rotor.

To operate the progressive cavity pump, an electric motor turns a drive assembly. Sucker rods attached to the motor above ground and to the pump at the bottom of the well. As these rods are turned by the motor at the top of the assembly, they transmit torque along their length. This torque results in a twisting of the rod string which stores a significant amount of energy. The system begins to move fluid when the torque provided to the pump is sufficient to overcome friction and viscosity. Pumping continues until shut down or until a problem arises.

Problem Statement

This type of drive assembly causes a problem in the event of shutting down the well, or a pump or motor failure. When a progressive cavity pump fails, it seizes up, the motor continues to apply torque to the suck rods, causing torque to build up in the drive string until the motor drops offline. This sudden release of positive torque will allow the drive string to unwind at an uncontrolled rate, inflicting significant damage to the assembly, possible injury to people, and may require the complete system to be replaced. If the motor fails due to other cause, the system will still have a large amount of potential energy stored in the sucker rods that will again be released when the motor stops providing torque. To deal with the problem, controlling the motion of the drive head/string is needed that will allow for the safe release if the stored energy while keeping the rod strings from uncoupling and unscrewing themselves from the threaded joints.

The uncontrolled backlash of the drive string is not the only problem. Generally, the pump will fail if there is not sufficient lubrication (fluid) to operate. If the system can be shut off before the pump runs dry, the life of the system can be increased, and the pump will only fail due to normal wear. To deal with the problem, detecting when a down hole pump is running dry is needed. If successful in preventing dry failure, the pump will only fail after excessive wear, which can be detected by reduced production rate.

Colloquial Need and Information Collection

Design a device that can be retrofitted to progressive cavity pumps to prevent pump failure and drive rod unscrewing is a colloquial need; that is, no damage is allowed to occur to the drive head or to the continuity of the drive system including the drive motor, gear box, socker rods, and couplings. To be able to identify the real need that implies viable design solutions, information is collected from the field and the process of pump failure is reviewed in-depth. The drive shaft and couplings appear to be the weakest links in the drive train when it comes to backlash failure. It is not enough to simply keep the torque in the string below that of shearing failure. The torque release mechanism must limit the motion of the drive rods so that the reverse torque from backlash does not unscrew the assembly. This type of failure is catastrophic because it means the rod string and pump must be extracted from the well.

To solve this catastrophic failure problem, three interrelated tasks must be accomplished. First the forward rotation if the assembly must be stopped. Second the assembly must be held stationary preventing it from rotating. Third the torsion in the sucker rods must be slowly released. To eliminate the problem of dry pump failure, a method of determining the state and amount of fluid in the progressive cavity pump is needed. By determining if the fluid pumped is not a lubricating liquid or if the reservoir liquid level has dropped below the top of the pump, the pump can be stopped avoiding damage to the elastomeric stator.

Identification of Real Need

Having gathered the available information and gained an understanding of the design task leads to the development of the need statement. A need is the problem to be addressed at its most basic level. Once identified, the need is divided into primary functions. These primary functions are tasks which the design solution must perform to satisfy the need. By performing Abstraction through the given statements, the true need statement is, "Safely release the energy stored in the drive rod." Here the primary function is to 'release the energy stored in the drive rod' and the primary constraint is to ensure that the releasing of the stored energy does not harm people, cause physical damage or unscrew drive rod couplings. The term 'safely' suggests proactive control. An additional constraint is implied by "the energy stored in the drive rod," which indicates that it is the potential energy that is to be released at a controlled rate. This precludes any solutions that encompass managing other than the torsional strain energy.

The function structure for the progressive cavity pump shown in Fig. 2.7 is constructed by employing the transformer model illustrated in Fig. 2.5.

Each of the first level functions are decomposed into their respective lowest (2nd) level subfunctions as follows:

Need Statement: A method to safely release the energy stored in the drive rod.

1. Recognize pump failure
   1.1 Measure torsion in rods
       DP: strain in rods
       CR: yield strength of rod
   1.2 Measure drive motor torque
       CR: maximum allowable motor torque
       DP: electrical current to motor
2. Stop rod movement
   2.1. Decelerate forward movement of the drive rod assembly
       DP: power to be dissipated
       CR: rotational kinetic energy
   2.2. Prevent rod assembly from rotating
       DP: static torque in rods

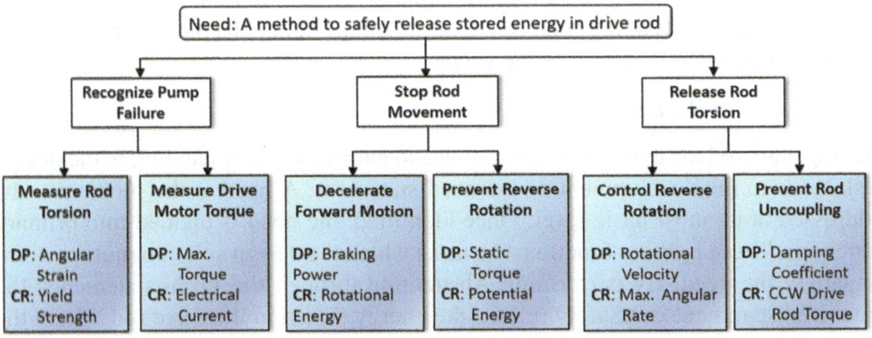

Fig. 2.7 Need statement and function structure of progressive cavity pump

CR: potential energy

3. Release rod torsion safely

   3.1. Control reverse rotation

      DP: maximum rotational velocity
      CR: maximum rotational rate allowed before damage occurs

   3.2. Prevent rod from uncoupling

      DP: damping
      CR: counter-clockwise (CCW) drive rod torque

The need statement and the accompanying function structure is realized and subsequently improved by questioning each of the functional requirements, constraints, and design parameters. In addition to gaining a better understanding of the design task, the function structure is also a starting point for formulating innovative design solutions.

2.4 Dust Storm Design

Semiconductor fabrication vacuum mainframes typically do not generate detrimental particles that affect the wafer die yield, but once these platforms are integrated with process chambers and operated in a manufacturing environment, processes and equipment such as O-rings and robot bearings create particles. These particles may cause a short in particle sensitive wafer circuitry. New vacuum mainframes, vacuum modules and sequences have been designed, however, access to existing process tools is technically involved and also extremely costly. An economical method and materials that would allow evaluation of new vacuum mainframes, vacuum modules and sequences without affecting the manufacturing process or inconveniencing the customer with iterative design development and validation is being sought. The method elaborated in the example creates wafer processing particles (geometry, chemical makeup, electrostatic, magnetic, and other properties similar) and induce these particles into a load-lock environment where mainframe hardware and process steps can be evaluated.

Identification of Real Need

1. A method to create wafer processing particles with similar geometric, electrostatic, magnetic, and chemical properties and induce these particles into a clean system.

In this need statement the problem is not clearly define

2. A method of inducing 65 nm and larger particles into a sanitary, low pressure environment with the capability of measuring the amount and size of particles as small as 65 nm.

There is a redundancy in stating "65 nm" twice

3. A method of inducing 65 nm and larger particles with similar magnetic, electrostatic, thermal and chemical properties to wafer processing particles into a sanitary, low-pressure environment and measuring the amount and size of the particles.

The types or particle properties does not need to be listed

4. A method of inducing 65–70 nm particles with similar properties to particles generated by wafer board processes into a low pressure, sanitary environment where the amount and size of the particles can be measured.

This need statement does not clearly define the environment in which the particles will be induced and the particles will not be measured in that environment

5. A method of inducing a controlled amount of 65–70 nm particles into a load-lock environment.

There is nothing in this need statement about the properties of these particles

6. A method of inducing a controlled amount of 65–70 nm particles of known properties into a working load-lock environment.

This need statement captures the scope of the problem and increases the number of solutions by forcing conceptual thinking

Once the perceived need is analyzed and the real need along with the constraints is identified through employing abstraction, a function structure is constructed in Fig. 2.8.

Design Specifications

Table 2.1 tabulates the design requirements for the functional and non-functional requirement are listed below:

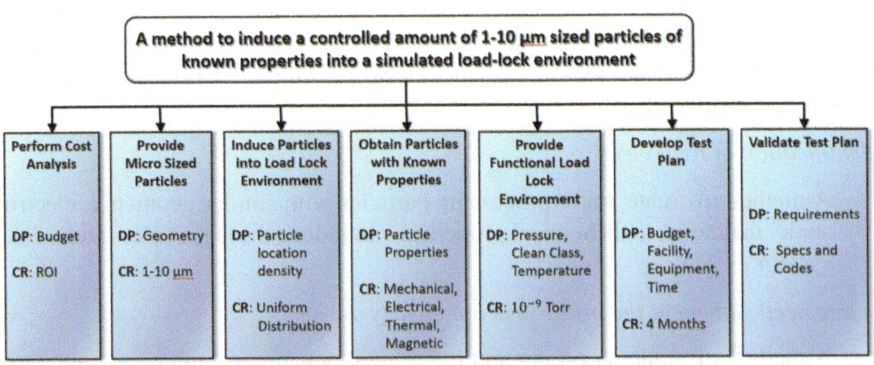

Fig. 2.8 Need statement and function structure for Dust Storm design

Table 2.1 Design specifications corresponding to the function structure

Function	Design parameter	Primary constraint	Comments
1-Provide a return on investment	Money	ROI of 10% or greater	A return on investment will be determined if the device is able to provide a higher chip yield without increasing capital or operational cost
2-Provide micro sized particles	Geometry	1–10 μm	The size of particle used will be determined by the best particle counter that can be obtained. The particles will either be the actual particles created or particles that have the same characteristics pending cost consideration
3-Induce particles into the load lock environment	Particle location density	Uniform or specific distribution	The particles must be induced into the system without affecting any other part of the wafer processing chamber
4-Obtain particles with similar properties to those found in a load lock environment	Particle properties	Mechanical, electrical, magnetic, thermal	Some particles that have been identified in the chamber are stainless steel, aluminum fluoride, silicon, and aluminum oxide
5-Provide functional Load Lock environment	Pressure, clean class and temperature	10^{-9} Torr	Must provide an accurate load lock environment
6-Develop test plan	Time, facilities, equipment, budget	4 months	A test plan will be developed where facilities, equipment, instrumentation will be recommended
7-Validate test method	Requirements	Meet requirements and specs	Before testing can begin, design concept must be validated using fluorescent particles and a black light for visual confirmation

2.5 Identification of Information and Resources Needs

Once the function structure is developed and functional requirements and design requirements are determined, a better understanding about requirements for a design solution is established. A properly performed need analysis enables the identification of the resource and knowledge needs required to pursue the design solution such as:

1. For the design to perform: Allocation of resources in energy, material, information, and users.
2. To manufacture the design:
    a. Allocation of resources: Energy and material needs.
    b. Recognition of information needs: Delivery of on-time information by specialists and information on the resources.
    c. Recognition of knowledge needs: Areas where breadths or depths of knowledge are required.
    d. Facility capabilities: manufacturing processes and space constraints.
    e. Identification of required personnel: Specialists in manufacturing, materials, manufacturing and marketing.
    f. Finance: allocation of finance for capital and running costs.
    g. Innovation: identification of areas that could be used for protection of intellectual property.
    h. Time constraints: Time to market, parts delivery schedule and time constraints on resources.

Prior to pursuing any solution, two things must be identified. One is the knowledge and resources that are available. The other is the ones that are not at hand but are needed in order to pursue the design. Earlier we have considered a design as a transformer. This transformation of resources can be thought of as occurring in two different aspects. The first aspect addresses the transformation of the desired set of inputs to the outputs in order to fulfill the functional needs of the design. The second aspect addresses the needs in producing the design. The first aspect enables the designer to perform the activities in the need analysis like the development of function structure and design requirements. The second aspect enables identification of informational and resource needs necessary to carry out the design tasks. Aspect 1 is usually the view of an engineer and aspect 2 is the view of a manager. Here both the views are considered and required in order to produce the design, as both are coupled with one another.

In the Dust Storm design, the development of the function structure and the design requirements addressed the first aspect. Considering the second aspect for the example, issues that must be addressed are:

- If the design solution is to be implemented, the manufacturing processes and materials must be determined accordingly to generate wafer processing particles of similar geometry, chemical makeup, electrostatic, magnetic, and all relevant properties.

- The materials could be nano powders of alumina, silica, iron oxide, and titanium that are of sizes ranging from 20 to 150 nm based on the constraints.
- The particles must be prepared according to ISO-14887 and the device must be certified. Thus, required information for certification must be gathered.
- The personnel required are to operate the device in class 100 clean room involving the use of high-speed vacuum pump, electrical supply and control, and safety regulation.
- The design team must be organized and the on-time delivery of information and input by specialists and suppliers.
- Capital cost and the time-to-validation must be established.
- Areas in the design or manufacturing of the device that require protection of intellectual property must be identified.

2.6 Planning for Innovation Opportunities

Need analysis not only facilitates innovation, but it also opens up opportunities for innovation. By keeping the function structure and design requirements solution independent, the solution space is broadened and kept wide open. Also, by the function structure breakdown, specific functions and features in the design that are candidates for innovation and patent protection can be readily identified. With novel features been identified early in the design process, measures can be taken to protect intellectual property and maintain competitive edge.

2.7 Discussion and Summary

Need analysis transforms a non-technical and solution specific, configurational design task to a true need. The functions, constraints and parameters that are inherent of the solution are derived from analyzing the true need through abstraction, critical parameter identification, and questioning. Need analysis fosters innovation by promoting objectivity to avoid fixation and solution dependency. Abstracting to extract the true need allows a comprehensive understanding of the scope and depth of the design task to be gained along with identifying the problem and solution domains. Powerful for driving the design process, need analysis is facilitated by being quantitative. Knowledge gained in identifying the real need along with the analysis outputs are carried over to the conceptual design stage. The outputs of the need analysis stage are crucial for allocating and managing resources including people, materials required, and manufacturing processes. A better understanding of the real need is also beneficial for utilizing experts for their expertise. Another prominent benefit a properly performed need analysis provides is identify areas in the design that may require patent protection.

References

Cross, N. (2000). *Engineering design methods: strategies for product design*. John Wiley & Sons.
Pahl, G., & Beitz, W. (1996). *Engineering design: a systematic approach*. Springer Verlag.

Chapter 3
Axiomatic Design: A Theory for Guiding Design Process

Axiomatic Design (AD) theory aims to establish a scientific design procedure for design. It develops design solutions to satisfy the customer requirement through the iterative process, where functions of a product and the design parameters are decomposed layer by layer. As a fundamental design theory of the innovative design thinking principles, AD is introduced briefly in this chapter.

3.1 Domains

AD theory divides the design world into four domains: customer domain, functional domain, physical domain and process domain. The relationship between domains is shown in Fig. 3.1. The four fields are arranged in order. The left field relative to the right field represents "the goal you want to achieve", and the right field relative to the left field represents "how to meet the goal of the left field".

The contents in customer domain are the customer's requirements or attributes of the product, process, system or material. Sometimes it is difficult to define the customer's requirements, or the requirements tend to be very vague. However, it is necessary to understand the requirements as much as possible. Working with the customer will be helpful.

In the functional domain, customer requirements are specified by functional requirements (FRs) and constraints (Cs). Functional requirements are the minimum set of independent requirements that completely represent the functional needs of the product, or software, organizations, systems, etc. According to the definition, when each functional requirement is established, it should be independent of any other functional requirements. Constraints (Cs) are the acceptable boundary of design, including two kinds of constraints: input constraints and system constraints. Input constraints are part of design specification, which are defined at the beginning of design. Input constraints are the stipulation of the design object, typically including size range, material cost, etc. The system constraints are derived from the design

W. Li et al., *Principles of Innovative Design Thinking*,
https://doi.org/10.1007/978-981-19-0485-1_3

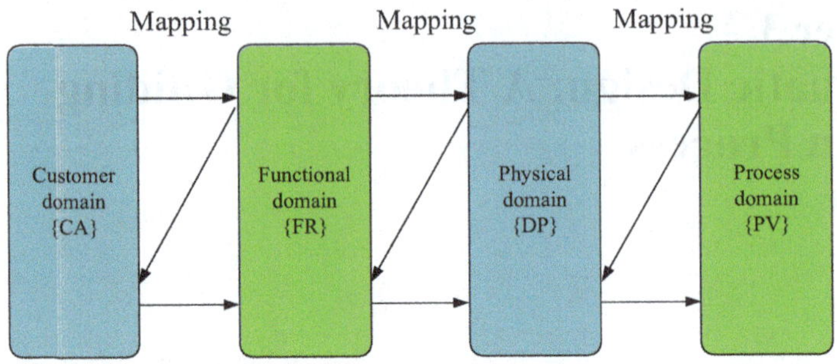

Fig. 3.1 Domains in AD

decisions, and all the high-level decisions play a constraint role for the low-level. In AD theory, constraints are not required to be independent of each other. When the number of constraints increases, the possibility of selecting the same FRs and DPs will increase. In this way, the design schemes will be more focused.

Design parameters (DPs) are the key physical variables in the physical domain, which are used to characterize the design and meet the specified functional requirements. When the functional domain consists of the functional requirements of the product, the corresponding design parameters can be the physical variables to meet the functional requirements. When the functional domain consists of the property of material requirements, the corresponding design parameters can be microstructure. When the functional domain consists of the functional requirements of software code, the corresponding design parameters can be input parameters, algorithms, modules, codes, etc. When the functional domain consists of the system, the corresponding design parameters can be equipment, or components, etc.

Process variables (PVs) are the key variables in the process domain that characterize the process that can generate the specified design parameters, or other equivalent term in the case of software design, etc.

3.2 Mapping Process

Functional requirements (FRs), design parameters (DPs) and process variables (PVs) must be decomposed until the design is completed. In particular, the decomposition should be completed by zigzagging between the domains instead of remaining in one domain. The decomposition process is proceeded layer by layer until the design reaches the final stage, generating a design solution that can be implemented. Through the decomposition process, the hierarchies of FRs, DPs, and PVs are established, which are the representation of the design architecture.

The mapping between customer domain (CA) and functional domain (FR) is to translate customer needs to functional requirements. During the mapping process, the functional requirements must be defined without ever thinking about something exists or what the design solution should be. If there was an idea of some existing design scheme, it will be designed according to this design scheme, then the design solution may be similar to the existing product, which will reduce the innovation of the design.

The decompositions of FRs and DPs are necessary to obtain the final design scheme. When the FRs are defined, the next step is to conceptualize a design solution. The conceptualization is during the mapping process going from the functional domain to the physical domain. After the overall design concept is generated by mapping, the DPs must be identified to complete the mapping process. The mapping process between domains can be expressed mathematically with the defined characteristic vectors.

FR vector in the functional domain is constituted by the set of functional requirements which defines the specific design goals, while the DP vector in the physical domain is constituted by the design parameters which are selected to satisfy the corresponding FRs. The relationship between FR vector and DP vector is expressed in Eq. (3.1).

$$\{\mathrm{FR}\} = [A]\{\mathrm{DP}\} \tag{3.1}$$

where $[A]$ is called the design matrix. For a design that has m FRs and n DPs, the design matrix is of the following form:

$$[A] = \begin{bmatrix} A_{11} & A_{12} & \cdots & A_{1n} \\ A_{21} & A_{22} & \cdots & A_{2n} \\ \vdots & \vdots & & \vdots \\ A_{m1} & A_{m2} & \cdots & A_{mn} \end{bmatrix} \tag{3.2}$$

Equation (3.1) can be written as:

$$\{d\mathrm{FR}\} = [A]\{d\mathrm{DP}\} \tag{3.3}$$

where the elements in $[A]$ are expressed by:

$$A_{ij} = \frac{\partial \mathrm{FR}_i}{\partial \mathrm{DP}_j} \tag{3.4}$$

Then functional requirement FR_i can be expressed as

$$\mathrm{FR}_i = \sum_{j=1}^{n} A_{ij}\mathrm{DP}_j \tag{3.5}$$

In the mapping process, there are several ways to define the design parameters.

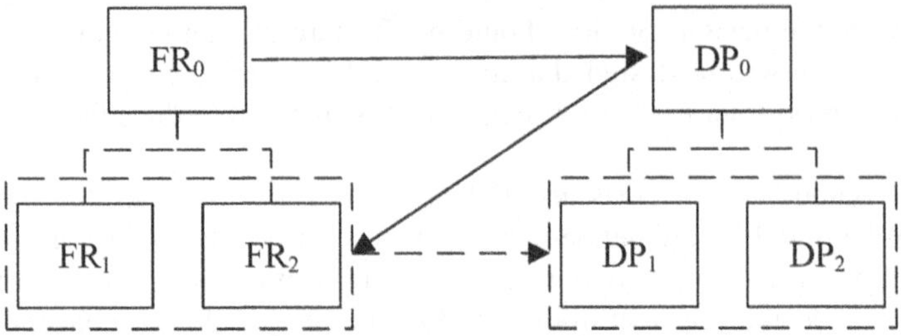

Fig. 3.2 The zigzag mapping process between domains

(1) Consider a specific DP to satisfy a specific FR, and repeat the process until the design is completed. It makes design process simple and convenient. However, in this way, there only be one design solution can be obtained finally. Moreover, if it is found in the decomposition of the lower level that the decomposition cannot continue due to improper selection of design parameters at the upper level, it is necessary to return to the upper level for re-decomposition, which will increase the iterations as well as the workload.
(2) Refer to the database to determine the DP. The database can help expand the design idea. However, since there are many categories and different organization forms in the database, the designer should have the corresponding knowledge and ability to select the proper design solution.
(3) Analogy with other cases. In this process, reverse engineering can be applied to copy someone else's good idea by examining an existing product. The method of scheme analogy enables designers to find a feasible design scheme in a short time. However, to a certain extent, this method is not conducive to the generation of innovative schemes.

From a FR in the functional domain, we go to the physical domain to conceptualized a design and determine its corresponding DP. When the FR and DP are defined completely at the high level, we go back to the functional domain to create the FRs at the lower level that collectively satisfies the higher-level FR. The decomposition process is pursued until all of the branches reach the final state and all the functional requirements can be satisfied. The zigzag mapping process to decompose FRs and DPs is shown in Fig. 3.2.

3.3 Axioms

AD theory normalizes design behavior with two axioms, namely "Independence Axiom" and "Information Axiom". The two axioms are based on the observation of physical laws, and have been confirmed in practical application. The design conforming to the Independence Axiom and Information Axiom has been improved in performance, robustness, reliability and practicability. At the same time, AD theory can be applied to analyze and solve problems for machines and systems that do not work well.

Axiom 1: The Independence Axiom. Maintain the independence of the functional requirements (FRs)

It states that when there are more than one functional requirement to be satisfied, the design solution must be such that each one of the FRs is satisfied without affecting the others. As a result, the selected design parameters must be able to satisfy the FRs and maintain their independence. It is noteworthy that the Independence Axiom requires that the function of the design be independent from each other, not the physical parts. Moreover, we must understand that the physical parts are not equal to the design parameters. To state the difference between physical parts and design parameters, we take the plastic bag for example.

Example 3.1 As shown in Fig. 3.3, a plastic bag is a whole part which has only one piece as the physical part. However, there are more than one FRs of the plastic bag: hold the items in it, easy to carry, cover (or show) the items in it and others. To satisfy the Independence Axiom, there must be more than one corresponding DPs: the body, the handles, the color, and etc.

In AD theory, the mapping process between the domains can be expressed mathematically as shown in Eqs. (3.1)–(3.5) The matrix $[A]$ has three forms: diagonal matrix, triangular matrix and other matrix types. In order to satisfy the Independence Axiom, the design matrix is either a diagonal matrix or a triangular matrix. When the design matrix is a diagonal matrix, it means that each functional requirement has a unique and definite design parameter. Such a design solution is called an

Fig. 3.3 A plastic bag

uncoupled design. When the design matrix is a triangular matrix, the independence of functional requirements can be guaranteed if and only if the design parameters are determined in an appropriate sequence, which is the arrangement sequence of DPs when the design matrix is a lower triangular matrix. Such a design is called a decoupled design. When the design matrix is a matrix other than diagonal matrix and triangular matrix, it means that the design is a coupled design. Uncoupled design and decoupled design are acceptable designs, while coupled design is not.

In order to get an acceptable design, it is necessary to make the number of DPs equal to the number of FRs. When the number of DPs is less than the number of FRs, a coupled design results or FRs will not be satisfied. When the number of DPs is more than the number of FRs, a redundant design or a coupled design will be produced. In an ideal design, the number of DPs is equal to the number of FRs, and the FRs are always independent from each other.

Axiom 2: The Information Axiom. Minimize the information content of the design

When there are several design solutions that satisfy a given set of FRs and all of them are acceptable in terms of the Independence Axiom. How to select the best one will be a problem. The Information Axiom provides a quantitative measure of the merits of a given design as well as the theoretical basis for design optimization and robust design.

The Information Axiom states that the design with the highest probability of success is the best design. The information content I_i for a given functional requirement FR_i is defined in terms of the probability P_i of satisfying FR_i.

$$I_i = \log_2 \frac{1}{P_i} = -\log_2 P_i \tag{3.6}$$

The logarithmic function is chosen to allow the information content to be additive when there are many FRs to be satisfied simultaneously. In the general case of n FRs, the information content for the entire system I_{sys} is

$$I_{\text{sys}} = -\log_2 P_{\{n\}} \tag{3.7}$$

where $P_{\{n\}}$ is the joint probability that all n FRs are satisfied. When all FRs independent statistically, as is the case for an uncoupled design, the $P_{\{n\}}$ is expressed as

$$P_{\{n\}} = \prod_{i=1}^{n} P_i \tag{3.8}$$

I_{sys} can be rewritten as

Fig. 3.4 The ranges for a FR

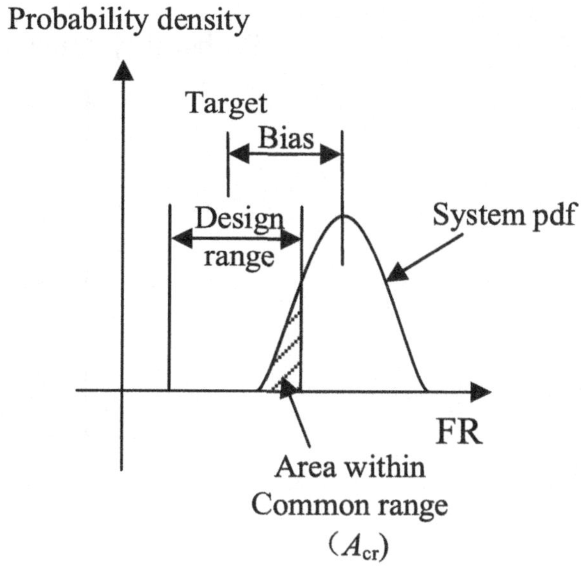

$$I_{sys} = \sum_{i=1}^{n} I_i = -\sum_{i=1}^{n} \log_2 P_i \tag{3.9}$$

Actually, the probability of success is governed by the intersection of the design range and the system range. The design range is defined by the designers to satisfy the given FRs. The system range is determined by the proposed design to satisfy the FRs. The overlap between the design range and system range is called the common range, which is the only region where the FR is satisfied. The ranges are shown in Fig. 3.4.

The information content may be expressed as

$$I = \log_2 \frac{1}{A_{cr}} \tag{3.10}$$

The ultimate goal of design is to minimize the information content as stated by Information Axiom. The relationship between ranges in an ideal design is shown in Fig. 3.5. To achieve the ultimate goal, there are two main methods: eliminating the bias and reducing the variance.

It is remarkable that the Information Axiom provide a powerful criterion for selecting the best set of DPs without the arbitrary weighting factors used in other decision-making theories. It is mainly because that the total information content will no longer represent the total probability if the information terms have been modified by multiplying with weighting factors before they are summed up. To prove this, we suppose each information content term has a weighting factor k_i, then the information content for the entire system I_{sys} in Eq. (3.9) is

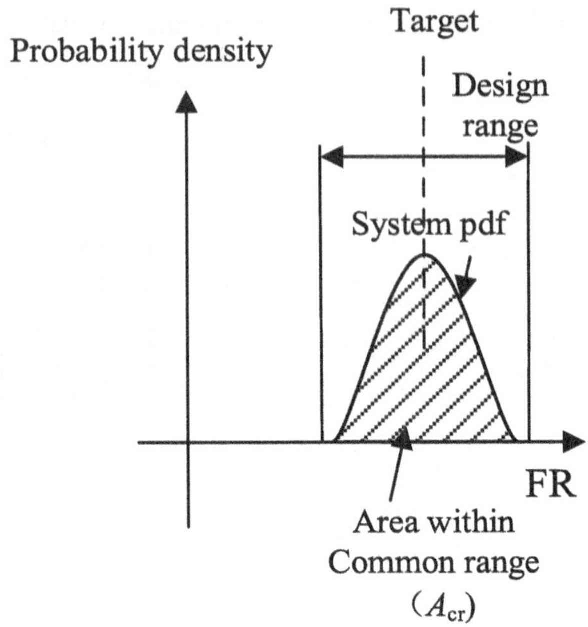

Fig. 3.5 An ideal design satisfying the Information Axiom

$$I_{sys} = \sum_{i=1}^{n} k_i I_i = -\sum_{i=1}^{n} k_i \log_2 P_i \tag{3.11}$$

The right-hand side of Eq. (3.11) is

$$-\sum_{i=1}^{n} k_i \log_2 P_i = -\sum_{i=1}^{n} \log_2 P_i^{k_i} = -\log_2\left(P_1^{k_1} P_2^{k_2} \cdots P_n^{k_n}\right) \tag{3.12}$$

Clearly, the content in the brackets is not equal to the total probability ($\prod_{i=1}^{n} P_i$), making the system information content cannot represent the total probability. It violates the definition that the information content is defined with the probability of success.

In case the designer thinks some FRs are very important, he or she can express the intention and the importance assigned to each FR with the design range, that is, the designer can redefine the design range with a tougher standard.

However, sometimes the design range is precisely specified while the designer still wants to emphasize the importance of some FR. For example, a functional requirement FR_i is so important to the designer that he or she wants it to be fully satisfied (or to some specified extent at least). In this situation, we suggest to list the specified information content for FR_i separately. The following example illustrate the point.

Example 3.2 Dr. Li wants to design a kind of table, he thinks that there are three important functional requirements that must be satisfied.

FR_1: The load-bearing capacity of the table must be enough, i.e., the bearing weight must be more than 300 kg.

FR_2: The table can be used by more than 5 persons at the same time, i.e., the area of the table must be larger than 1.2 m^2.

FR_3: The table can tolerate the temperature more than 1300 °C.

Now he has got three design schemes (Table 3.1).

Which design scheme is the best according to Information Axiom?

The probability distributions of the FRs in scheme A are shown as follows (Figs. 3.6, 3.7 and 3.8):

The probability distributions of the FRs in scheme B are shown as follows (Figs. 3.9, 3.10 and 3.11):

The probability distributions of the FRs in scheme C are shown as follows (Figs. 3.12, 3.13 and 3.14):

The information content for each scheme is calculated with Eq. (3.9). The results are as follows (Table 3.2).

Finding the optimal design scheme is to select the minimum value of I_{sys}:

$$I_{\text{sys(op)}} = \min I_{\text{sys}(i)} (i = 1, 2, 3) \tag{3.13}$$

Table 3.1 Three design schemes

Design schemes	FR_1 = The bearing weight (kg)	FR_2 = The area (m^2)	FR_3 = The temperature (°C)
A	310–350	0.8–1.0	1200–1800
B	280–350	1.0–1.5	1300–1900
C	400–500	1.5–2.0	1100–1700

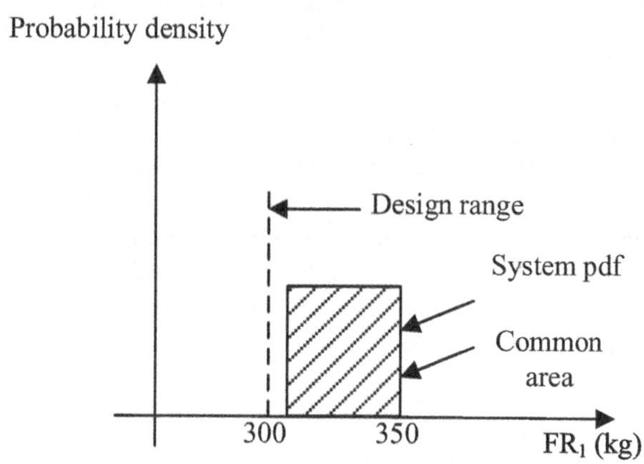

Fig. 3.6 The probability distributions of FR_1 in scheme A

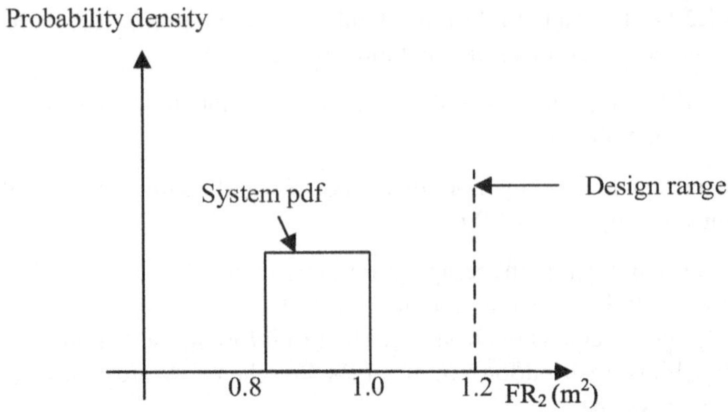

Fig. 3.7 The probability distributions of FR_2 in scheme A

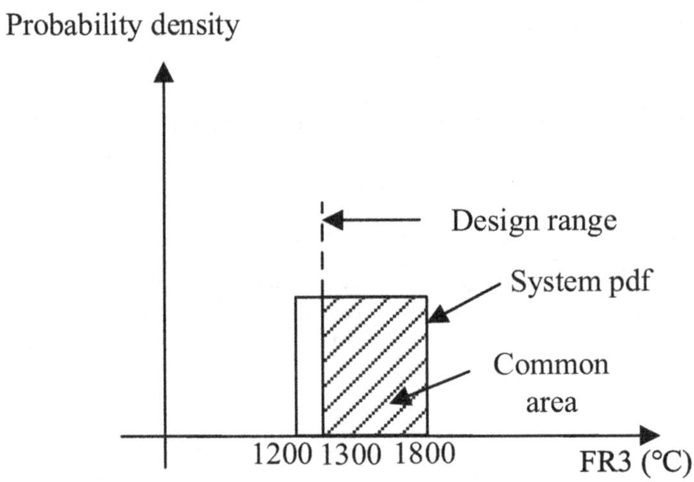

Fig. 3.8 The probability distributions of FR_3 in scheme A

In this example, the information content is the minimum in design scheme C, making C the best choice.

Now, if the designer thinks the FR_2 is more important to him, he can express the importance by mean of the specification of design range instead of a weighting factor. For example, he can specify the length or width of table.

In another situation, Dr. Li wants the table to tolerate the temperature more than 1300 °C in any case. The design range remains the same while FR_3 must be fully satisfied. Finding the optimal design scheme is to solve Eq. (3.13) under the condition:

$$I_3 = 0 \quad (3.14)$$

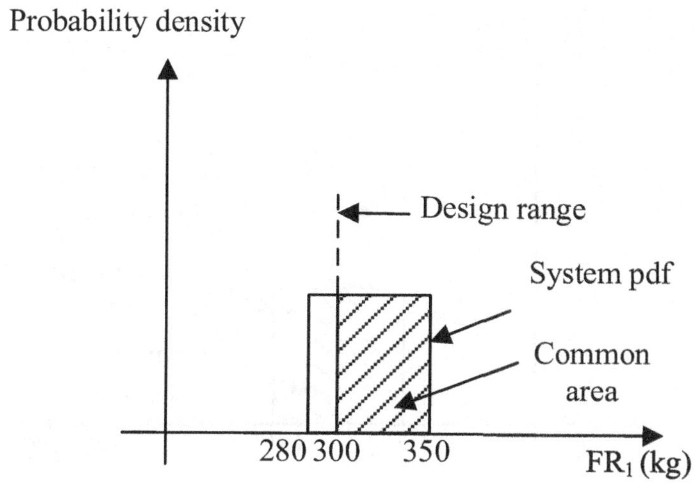

Fig. 3.9 The probability distributions of FR_1 in scheme B

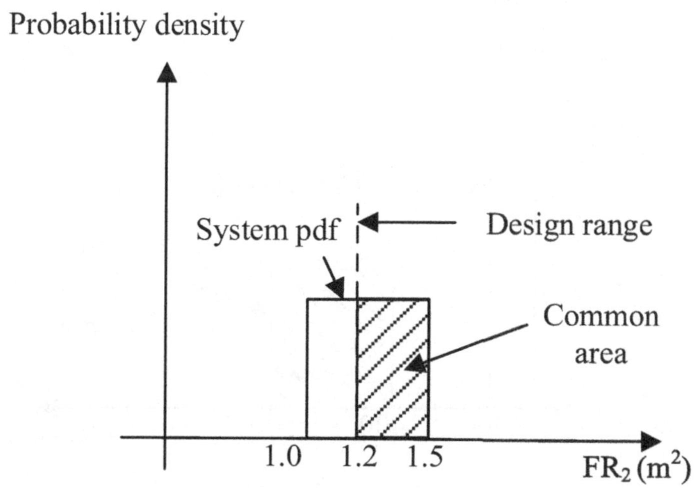

Fig. 3.10 The probability distributions of FR_2 in scheme B

In this case, the optimal design scheme is B.

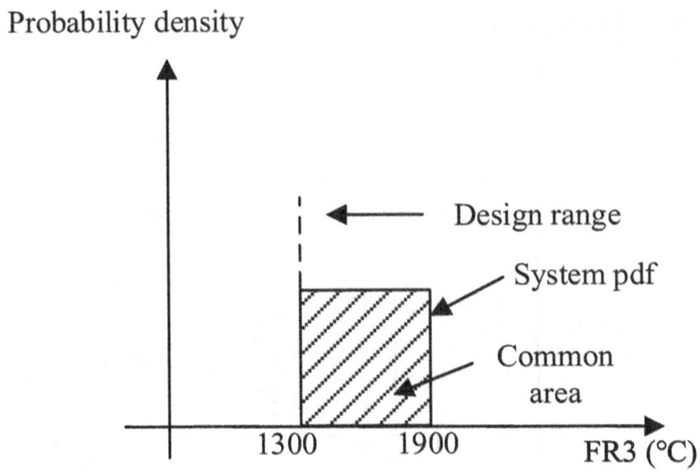

Fig. 3.11 The probability distributions of FR_3 in scheme B

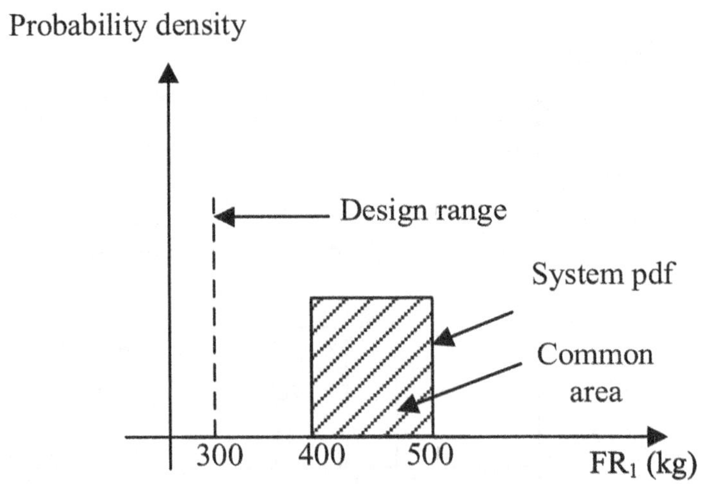

Fig. 3.12 The probability distributions of FR_1 in scheme C

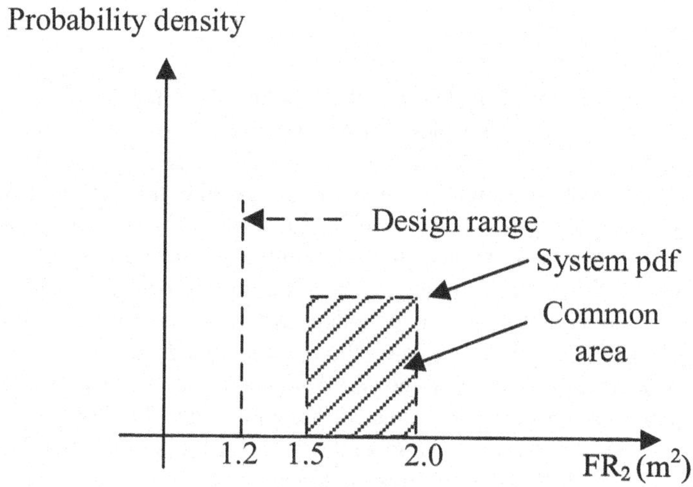

Fig. 3.13 The probability distributions of FR_2 in scheme C

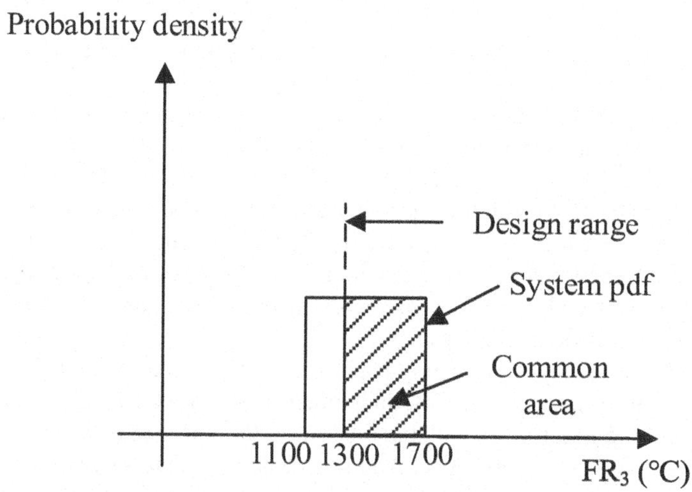

Fig. 3.14 The probability distributions of FR_3 in scheme C

Table 3.2 The information content for each scheme

Design scheme	I_1	I_2	I_3	I_{sys}
A	0	∞	0.263	∞
B	0.485	0.678	0	1.163
C	0	0	0.585	0.585

3.4 Coupling

Independence Axiom makes it clear that the FRs must be satisfied by a correct set of DPs without affecting the other FRs. The Independence Axiom is especially important in the mapping process. When FRs are defined, a series of possible solutions can be found and evaluated. In this evaluation process, the Independence Axiom can help designers judge whether the solution is a good one. If the Independence Axiom is obviously violated, it is known that the solution is not a good design without further analysis. According to Independence Axiom, the design matrix (Eq. (3.2)) must be either a diagonal matrix (uncoupled design) or a triangular matrix (decoupled design). If the design matrix is in other forms, the design is a coupled design, which is not acceptable. Based on the two design axioms, AD derives a number of corollaries and theorems relevant to the coupling problem (Suh 2001, 60–64).

Theorem 4 (Ideal Design): In an ideal design, the number of DPs is equal to the number of FRs and the FRs are always maintained to be independent of each other.

Theorem 8 (Independence and Design Range): A design is an uncoupled design when the designer-specified range is greater than $\left(\sum_{\substack{i \neq j \\ j=1}}^{n} \frac{\partial FR_i}{\partial DP_j} \Delta DP_j\right)$, in which case the non-diagonal elements of the design matrix can be neglected from the design consideration.

Theorem 20 (Design Range and Coupling): If the design ranges of an uncoupled or decoupled design are tightened, they may become coupled designs.

Brown (2006) has divided the coupling into three kinds: FR-FR, DP-FR, and DP-DP coupling, and FR-DP coupling is not considered because a function categorically cannot interfere with a design parameter. FR-FR coupling occurs when one FR overlaps and therefore interferes with another. DP-FR couplings occurs when a DP is selected to fulfill an FR, while it interacts with other FRs. DP-DP coupling usually happens during physical integration.

Although the concept of coupling is put forward in AD and the design matrix is defined to identify the coupling, the judgement is still very vague in practice. The elements in design matrix are obtained based on the knowledge and experience of designers, which is very subjective. And it is difficult to determine whether there is an "influence" relationship between two FRs. In order to clarify the connotation and significance of the Independence Axiom, this book summarizes ten rules to facilitate the understanding and application of the Independence Axiom.

Rule 1: *The "independence" is not an absolute independence, instead, it is conditional and delimited.*

From the perspective of system theory, everything is related, especially in the same system. It is impossible for a subsystem to be absolutely independent from other ones and not be affected completely. So, there must be some conditions to the Independence Axiom. According to Independence Axiom, when the change of a design parameter (DP_i) cannot lead to any change of the other functional requirements, the corresponding functional requirement (FR_i) is considered to be independent from other FRs. Herein, the change of DP_i should be within a certain range, which may be a well-known range or an artificially limited range. If the range is not limited and DP_i is allowed to change to some extreme state, any functional requirement could be affected. However, in the application process, people tend to default that the change of DP_i is limited to a reasonable range. Therefore, it may be hard to realize the importance of the "range" to the Independence Axiom.

Rule 2: *In most cases, uncoupled design cannot be obtained. Instead, the design schemes are usually decoupled design with triangular matrix.*

Most iterations usually result in decoupled design. It is because there are many interrelationships between subsystems that are difficult to eliminate. For example, due to the DP-DP coupling caused by physical integration, one component needs to have physical contact with another component when it's working. For instance, the wheel of a bicycle must be correctly assembled on the body to work.

Another example is the coupling caused by the sequence of functions, which is usually inherent and inevitable, especially in the manufacturing industry and other fields. The needs of specific operation sequence and workflow make this kind of coupling more common. Therefore, many engineering designs are decoupled and they are also acceptable according to AD theory.

Rule 3: *Decoupling is not necessary for some acceptable coupling.*

Such coupling includes:

(1) Necessary physical relations in the physical integration.
(2) Suppose there is coupling between FR_1 and FR_2, while the change of FR_1 caused by DP_2 is within an acceptable tolerance, that is:

$$CL_1 \leq \frac{\partial FR_1}{\partial DP_2} \Delta DP_2 \leq CU_1 \tag{3.15}$$

where CL_1 and CL_2 represent the upper and lower bounds of FR_1 respectively. If the change of FR_1 caused by DP_2 is within its acceptable range, FR_1 is considered to be not sensitive to DP_2 and the effect of DP_2 can be ignored.

(3) The decoupled design with triangular matrix. Of cause, it would be better to eliminate the coupling if possible. But if the decoupling cost is too high or the improvement is not significant, we can choose to ignore the coupling since the decoupled design is also acceptable.

Rule 4: *The values of the elements in design matrix usually depend on the designers' consensus.*

Sometimes, the consensus is not always consistent with the actual situation. It is related to the personal ability and the interaction in a design team. Not all of the designers can accurately evaluate the interaction between DPs and FRs. The accuracy of judgment depends on the knowledge accumulated within the team. If it is difficult for the designers to determine the effect of some DP to a FR, more knowledge even the database is needed.

Rule 5: *It is difficult to establish the coupling relationship during the concept design process, due to the uncertainty of the specific components.*

According to AD theory, we must write down the design equation (Eq. (3.1) at each level of decomposition to ensure we have made the right design decision. When the Independence Axiom is violated by design decisions made, we should go back and redesign rather than proceeding with a flawed design. However, when following the zigzag mapping process of AD, the decomposition of the first layers is still in the conceptual design decomposition stage, where the specific design parameters are not determined yet. Therefore, it is difficult to tell whether there is any coupling problem.

Rule 6: *Designers tend to select a specific solution for each design parameter when applying the AD axioms.*

According to Rule 5, conceptual design decomposition will bring uncertainty, which will make it difficult to determine the values in design matrix. Meanwhile, the design equation is required to be generated at each level, which results in a contradiction problem. In practical, the designers tend to find a relatively specific design parameter to meet the corresponding functional requirement.

For example, the specific object of "diesel engine" is used instead of the abstract expression of "energy conversion device" to meet the function of "generating energy". This tendency could make the solution space converge rapidly, and a specific design scheme would be obtained finally. But at the same time, the divergence of the solution space could be limited as well as the generation of various design schemes.

Rule 7: *It is better to check the effect of DP on each FR one by one to complete the design matrix.*

This rule is to ensure that every element in design matrix is taken into account. But when there are many DPs and FRs, the workload of checking one by one will be very heavy. In this case, we can turn to some appropriate algorithm or interaction database, and use computer for processing.

Rule 8: *When there is overlap between FRs, it must be decoupled.*

The overlapping of FRs is due to the fact that the function requirements are not mutually exclusive during the decomposition process. The overlap between FRs can be judged by establishing FR-FR matrix. If there exist some non-diagonal elements of the matrix that are not zero, there is FR-FR coupling. And the FR-FR matrices of this kind of coupling are symmetric. The design with FR-FR coupling must not be uncoupled or decoupled. The FR-FR matrix of uncoupled or decoupled design is diagonal matrix.

The only way to decouple such coupling is to make FRs mutually exclusive. The overlapping between FRs may occurs due to the fact that some constrains are misjudged as functions. Cost, time, weight, risk, etc. are some common constraints. The way to judge whether a variable is a constraint or a function is to determine whether it needs a DP to satisfy it. If a DP is needed, it means the variable is a functional requirement, and the coupling should be eliminated. Otherwise, it's a constraint and should be removed from the functional domain. In addition, when the change of a DP has similar effects on two or more FRs, it is likely that the two or more FRs are parent-and-child relationship instead of sibling relationship.

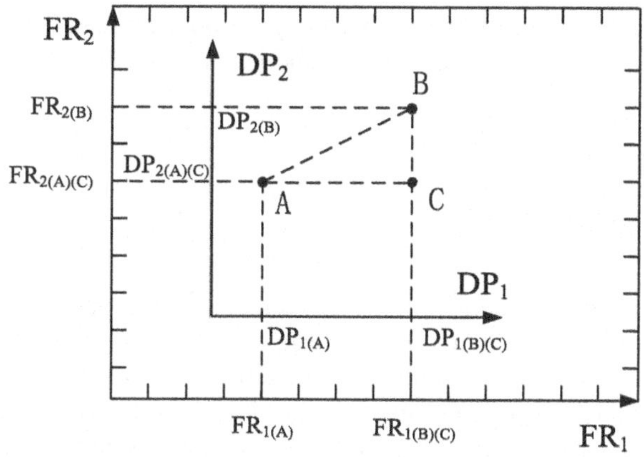

Fig. 3.15 FRs and DPs coinciding with Independence Axiom

Rule 9: *The Independence Axiom can be expressed by mathematical vector graph. When FRs are independent, DPs should be orthogonal to each other, while a DP and its corresponding FR should be in parallel relationship.*

Taking the design of two functional requirements and two design parameters as an example, the relationship diagram in accordance with the Independence Axiom is shown in Fig. 3.15.

The vertical relationship of FRs is set at the time of functional requirements decomposition. If there is overlap between FRs due to incorrect decomposition, refer to rule 8 for decoupling. If FRs are independent, it is shown that when the state of FR_1 is required to change from point A to C, only DP_1 needs to be changed from $DP_{1(A)}$ to $DP_{1(C)}$ without affecting FR_2. Similarly, when the state of FR_2 is changed from point C to B, only DP_2 needs to be changed from $DP_{2(C)}$ to $DP_{2(B)}$ without affecting FR_1. The advantage of function independence is also reflected, that is, when the function state of the whole system changes from A to B, it only needs to change $DP_{1(A)}$ to $DP_{1(B)}$ and $DP_{2(C)}$ to $DP_{2(B)}$ respectively, without iteration. And it won't be affected by the design sequence.

Rule 10: *The changes should be considered when the DPs work to satisfy FRs rather than in the design process.*

According to Rule 9, the changes of the FR and DP are the key parameters to be considered for the Independence Axiom. Thus, when and how the changes are considered are important and should be analyzed and made clear. It is known that all the components of a product are assembled together and there are lots of interactions and relationships between them, such as the assembling relations, the relationships among shapes and sizes. When making decision on determining the design components in design process, if one DP is changed, it may affect lots of other DPs according to the interactions and relationships. Obviously, it's not a proper time to consider the Independence Axiom.

Independence Axiom should be considered when the DPs are determined already and the changes are from one state to another when DPs work to satisfy the corresponding FRs.

3.5 Constraints

In the design process, it requires not only to realize the functions with the design parameters, but also to standardize the design with constraints (Cs). The design objectives always need to be within constrains. In some complex systems, the functional requirements cannot fully express the design requirements, and constrains are needed to provide acceptable boundaries for design solutions. The differences between constrains and functional requirements are as follows (Suh, 2001):

(1) FRs need to be independent of each other, while constrains don't have to be.
(2) FRs must be implemented with corresponding DPs, while constrains are not.
(3) FRs focus on the completion of actions, while constrains are the description of "non-functional" requirements in design, focusing more on the performance.

Function is used to describe what the product can achieves, which represents the fundamental significance of the product. Constraints are applied to ensure the product has good performance, which is the basic condition for the product to meet the needs of customers and capture the market. Taking the agricultural machinery as an example. China's agricultural machinery has basically completed its functions, but its performance still needs to be improved. With the further improvement of the function of agricultural machinery products, the design parameters increase rapidly, and the development of mechanical and electrical integration makes the potential failure of agricultural machinery products more and more likely. At present, there are many kinds of agricultural machinery products in China, with powerful functions and complex structures, but the reliability is poor and the average time between failures is short, which has become a major obstacle to the development of agricultural machinery in China. The reliability of agricultural machinery is one of the constraints in design. Reliability design is a system engineering, involving many design parameters. However, when the designer cannot fully understand and reflect the reliability constraint in the design, the performance of agricultural machinery products appears to be poor. In addition to reliability, the common constraints include time constraint, cost constraint, defective rate constraint, manufacturability constraint, weight constraint, volume constraint, precision constraint and so on, which together determine the performance of products. In order to analyze the constraints systematically, we first classify them.

Constraints can be divided into input constraints and system constraints according to the way they are generated. Input constraints are the limits of the whole design objectives, and all design schemes must meet these constraints. This kind of constrains include cost, energy consumption, etc.; while system constraints are the result of a series of design decisions in the design. For example, choose the motor as the power source, then subsequent decisions need to be based on this, and a series of constraints may be caused, such as endurance time.

According to the degree of the influence, constraints can be divided into main constraints and auxiliary constraints. The main constraint plays a decisive role in the mapping process or structure, which can greatly affect the result of mapping and the selection of structure. The auxiliary constraint's influence is relatively less than main constraint, but still essential.

Constraints can also be divided into quantitative constraints and qualitative constraints according to their properties. Quantitative constraints can be expressed by numerical values, and qualitative constraints can only be expressed by fuzzy words instead.

Different constraints have different solutions. For the constraints that can be expressed quantitatively, the relevant equations of constraints can be added in the calculation process, and the design scheme meeting the conditions can be selected. Take the cost constraint for example, the manufacturing cost is required to be less than a certain value. All manufacturing costs can be added and compared with the allowable value. For qualitative constraints, such as the high reliability of machine operation, the reliability of the whole machine can only be verified after the design is completed. The optimal design scheme can be selected by comparing the reliabilities of multiple design schemes.

In order to consider the constraints in the design process, Zhang et al. (2008) proposed an AD theoretical design method to solve design problems under constraint conditions, which is mainly divided into two steps. In the first step only the functional requirements are considered while the constraints are not. In this way the preliminary design scheme is generated. The second step is to improve the scheme according to the constraints. If there are multiple constraints, each constraint is considered to gradually improve the scheme. This method considers functions and constraints separately, which can simplify the problem to a certain extent, and is suitable for the case with simple constraints and functions.

When the constraints involve multiple sub functions and corresponding design parameters, it is likely that a series of design parameters need to be modified in the second step of improving the scheme by using the above method, which will lead to the failure of the whole preliminary design scheme and the redesign is required. Taking the tractor design as an example, the "minimum pollution" is taken as a constraint. If the initial design scheme is completed by zigzag mapping decomposition of AD theory without considering the constraints, the designer is likely to choose the commonly used diesel engine when selecting the power source. During the decomposition of FRs and DPs, the diesel engine will be the basis for matching to get the transmission system, control system and other subsystems suitable for the diesel engine. After the preliminary design scheme is completed, the "minimum pollution" constraint is used to improve the design scheme. It will be found that the diesel engine is the largest source of pollution. In order to reduce the pollution, it is necessary to replace the diesel engine with the motor, and the design schemes of the subsystems matching with the diesel engine in the original scheme will also be overturned and redesigned, which will greatly increase the design workload, resulting in a waste of time and resources.

Therefore, designers need to consider constraints in all levels of design, that is, the solution process of constraints should be reflected in the whole design process. Because AD theory does not form a detailed and systematic description model of functional requirements and design parameters, the decomposition and embodiment of constraints could not be easy to achieve.

3.6 Discussion and Summary

In this chapter, the AD theory is introduced along with the fundamental concepts of domains, mapping process, design axioms, coupling and constrains. Four domains of customer domain, functional domain, physical domain and process domain are defined to describe the design world. Mapping between domains is a zigzagging iterative process, through which the decomposition is completed and the design equations as well as design matrices are generated.

Independence Axiom states that the FRs must be maintained independent to each other by selecting the appropriate DPs. Information Axiom is used to ensure the information content of the design is minimum, where the information content is defined by the probability of success to satisfy the FRs.

According to the Independence Axiom, the concept of "coupling" is defined. Design schemes are classified into uncoupled design, decoupled design and coupled design. The corresponding design matrices are diagonal matrix, triangular matrix and other matrix types. What we want to achieve is to make the coupled design uncoupled or decoupled. Furthermore, ten rules were summarized by the authors in this chapter to help designers understand and implement the Independence Axiom.

In addition, another important concept of constrain is introduced, which provides acceptable boundaries for design solutions. The differences between constrains and functional requirements are discussed. The constrains are analyzed and classified to play a better role in the design process.

References

Brown, C. A. (2006). Kinds of coupling and approaches to deal with them. *Proceedings of 4th International Conference on Axiomatic Design (ICAD2006)*, Firenze, pp. 1–5.

Suh, N. P. (2001). *Axiomatic design: Advances and applications*. Oxford University Press.

Zhang, L., & Cao, J. (2008). A study on axiomatic designing method under constraint conditions. *Journal of Machine Design*.

Chapter 4
Extenics: A Methodology for Solving Wicked Problems

Wicked problems are a significant topic for the problem-solving endeavor in design thinking. Tame problems are well defined and can be resolved, whereas the solution for a wicked problem can only be better or worse, not right or wrong. Wicked problems are complex, indeterminate, and ill-defined problems because they have the characteristics of being incomplete, shifting, contradicting and interdepended information, which is difficult to gather. Buchanan (Buchanan, 1992) argued that design problems are wicked for the following reason: "*Design problems are "indeterminate" and "wicked" because design has no special subject matter of its own apart from what a designer conceives it to be. The subject matter of design is potentially universal in scope, because design thinking may be applied to any area of human experience. But in the process of application the designer must discover or invent a particular subject out of the problems and issues of specific circumstances.* " Cross (Cross, 2006) found that designers consider all design problems assuming they were wicked problems. In other words, wicked problems are fundamental assumptions within the realm and operation of design thinking.

4.1 Wicked Problems

The notion of wicked problems was first developed in 1973 by design theorists Horst W. J. Rittel and Melvin M. Webber, professors of design and urban planning, in the planning literature to describe social problems that did not correspond well to the accepted and practiced models of policy analysis. These problems involve multiple possible causes and coupled interrelations that are complex, difficult to define, and inherently impossible to address properly. Unlike tame problems for which a quick solution is oftentimes either available or easily conceived, wicked problems are complicated, interconnected, or simply too large in scope and degree of complexity to fix. While wicked problems such as hunger, poverty, terrorism, and sustainability may seem too complex to be solvable, employing design thinking provides the best

W. Li et al., *Principles of Innovative Design Thinking*,
https://doi.org/10.1007/978-981-19-0485-1_4

approach to a solution. According to Rittel and Webber (Rittel & Webber, 1973) the concept of wicked problems is defined by ten common characteristics as follows:

1. Wicked problems lack a definitive formula.
2. There is no stopping rule for determining when a solution has been found.
3. Good or bad solutions rather than true or false solutions.
4. Each wicked problem is unique to itself.
5. Solutions are not immediate and cannot be tested.
6. There is no trial and error when finding solutions, only implementation.
7. Wicked problems are interconnected or symptoms of other problems.
8. Wicked problems have more than one explanation.
9. There is no exhaustive list of possible solutions.
10. Those who try to solve wicked problems have no right to be wrong in that they are responsible for outcomes that result from the actions they take.

Wicked problems differ from other problems in three specific regards (Weber & Khademiam, 2008). They are:

1. Unstructured in that it is difficult to sort out causes and effects and little consensus in identifying problems and solutions,
2. Cross-cutting in that they have many overlapping stakeholders with different perspectives on the problem, and
3. Relentless and dynamic. They cannot be solved decisively with conclusive answers.

It is commonly believed that while it is not possible to find an elegant solution, nonetheless, wicked problems can be mitigated, not solved, through adopting an approach underlined with empathy, abductive reasoning, and rapid prototyping, among others. To channel mitigation efforts in desirable directions, people who are charged with mitigating the negative consequences of wicked problems are required to think like a designer does and apply methodical, rigorous iteration focused on the various properties of the problem. Interdisciplinary collaboration engaging a broader knowledge of science, economics, statistics, technology, psychology, and politics is necessary to effectuate solutions. So is employing abductive reasoning in addition to deductive and inductive approaches.

Abductive reasoning begins with making observations and progresses to drawing up potential justifications to clarify and explain the observations. While deductive reasoning seeks certainty and inductive reasoning enumerates uncertainty, abductive reasoning endeavors to generate meaning subject to uncertainty and ambiguity. Mitigating wicked problems requires changing the questions, managing uncertainty, and facilitating resilience. It does not resolve the problem but instead drives to a desired future state. Measures conceived to address wicked problems are never one and done. Customary approach with solution-proposing at its core is not effective. Transitioning to a process that defines a desired state is therefore essential. Solutions then balance and reconcile tradeoffs while considering self-interests of the stakeholders, indicating the pivotal roles played by collaboration and interdisciplinary approaches. Defining a wicked problems framework with adaptive management is key to differentiating good solutions from bad ones, not true or false.

The evolving and dynamic nature of wicked problems constantly introduces new constraints and interactions among different stakeholders, rendering orderly problem-solving approach not feasible. The process for finding wicked problem solution is not linear but rather requiring reflecting forward and backward with the goal of perceiving the problem from a new perspective; that is, reframing the problem through abstraction. Reframing is to probe the problem and attack the solution through rephrasing, to question what seems apparent, to determine if the problem is a symptom of another problem, so as to allow the true need to be identified and core issues be discovered. This is often done by reframing the problem to a higher level by making it more general. To address design problem with wicked property, the designer must frame the challenging design situation by articulating its scope and boundary, objects and their relations, and the logic for guiding subsequent design steps.

Novel designers are more knowledgeable of what is truly needed than they know how to communicate the need in precise terms. They reflect and retrieve salient knowledge while they are in action designing. Reflecting on the action taken and process followed is also an inherent part of the activity in which knowing-in-practice is demonstrated. When abductive reasoning is invoked, the implicit knowledge is utilized in practice. This is lucidly described by Schön (1983) as a reflective practice: *"In a good process of design, this conversation with the situation is reflective. In answer to the situation's back-talk, the designer reflects in action on the construction of the problem, the strategies of action, or the model of the phenomena, which have been implicit in his moves."* When encountering unexpected challenges experienced designers draw upon previous experiences to deploy inferred knowledge and reflection. Practicing reflection is therefore central to tackling problems with wicked properties.

4.2 Abductive Reasoning

We form inferences about the world through reasoning. Reason is the capacity for consciously making sense of things, applying logic, establishing and verifying facts, and changing or justifying practices, institutions, and beliefs based on logic and evidence. Inductive reasoning is a specific-to-general form of reasoning seeking to generalize based on specific instances whereas abductive reasoning is a specific-to-general form of reasoning that specifically looks at cause and effect.

"Inductive reasoning makes broad inferences from specific cases or observations. In this process of reasoning, general assertions are made based on specific pieces of evidence. Scientists use inductive reasoning to create theories and hypotheses. An example of inductive reasoning is, "The sun has risen every morning so far; therefore, the sun rises every morning." Inductive reasoning is more practical to the real-world because it does not rely on a known claim; however, for this same reason, inductive reasoning can lead to faulty conclusions.

Abductive reasoning is based on creating and testing hypotheses using the best information available. Abductive reasoning is used in a person's daily decision making because it works with whatever information is present—even if it is incomplete information. Essentially, this type of reasoning involves making educated guesses about the unknowable from observed phenomena. Examples of abductive reasoning include a doctor making a diagnosis based on test results and a jury using evidence to pass judgment on a case: in both scenarios, there is not a 100% guarantee of correctness—just the best guess based on the available evidence.

The difference between abductive reasoning and inductive reasoning is a subtle one; both use evidence to form guesses that are likely, but not guaranteed, to be true. However, abductive reasoning looks for cause-and-effect relationships, while induction seeks to determine general rules."—https://courses.lumenlearning.com/boundless-psychology/chapter/reasoning-and-inference/.

As opposed to deduction which is the logic of necessity and induction which is the logic of probability, abduction is the logic of possibility. Deduction is going from general to specific and induction is going from specific to general, whereas abduction is reasoning between analyzing evidence and data and engaging intuitive thinking through propositions. Because design is an iterative process iterating between the concept and configuration domains (see the Concept-Configuration Iteration Model in Chap. 1) employing testing solution propositions and qualified guesses using empirical observations such as visualization and rapid prototype, the reasoning of design thinking is fundamentally abductive.

As opposed to inductive reasoning which begins with observations that are specific and limited in scope and proceeds to a generalized conclusion based on accumulated evidence, abductive reasoning starts with making partial and incomplete observations and progresses to form logical explanations for the observations. While there is no absolute certainty about the outcome, given that the observations are incomplete and additional information may be overlooked, abductive reasoning allows the designer to make his best effort in generating good solution. While characterized by lack of completeness, either in evidence or in explanation, or both, the abductive process is widely credited for playing an essential role in creativity and innovation.

Design as an activity involving multiple stages is to generate solutions that are fully justified in the context for which they are intended. With no exception all design solutions are eventually evaluated to determine their values for being contextual meaningful. Design thinking seeks to embody contextual meaning which is also evident from the ubiquitous use of empathy, human-centered, co-creation, and participatory design in the literature. Abductive reasoning is performed in design thinking to realize relevant solutions that are of quality and meeting specific need and requirement. Evaluation criteria are defined in the realm of design thinking to ensure that the corresponding contextual meaning is exemplified by the design.

Reframing a design problem is an iterative process of learning in design thinking exploring abductive reasoning and abstraction for articulating the real need and generating knowledge. This learning process involves iterating through the concept and configuration domains engaging inspiration, ideation, embodiment, and prototyping. Reframing is conceptually repositioning the problem and reshaping the articulation to foster new innovative solution ideas. Reframing is also performed within semantics where the meaning of an object is reshaped (transformed) though association, derivation, or implication. Extenics, the theory of design thinking to be introduced in the following sections, provides a structured framework for exploiting linguistics to drive the evolutional changes of meaning and form.

4.3 Theoretical Framework of Extencis

Extenics was first formulated by Chinese scholar Cai (1999). As an emerging multidisciplinary theory with formalized, logical and mathematics characteristics, Extenics explores the extension and derivation of matter and object, and the formulation of rule and method of innovation (Cai, 1998). The logical cells of Extenics are matter-element, affair-element and relation-element (generally called basic-element). Extenics researches the extensibility of basic-elements, rules of transformation and calculation, the establishment of extensible models out of mathematics models for problems whose requirements are conflicting and irreconcilable, and the process of resolving the contradiction problems (Cai, 1998; Cai & Shi, 2006; Cai et al., 2001, 2005). Contradiction in Extenics means that specific goals cannot be reached under current conditions.

The theoretical basis of Extencis is extension theory, including basic-element theory, extension set theory and extension logic constitute the extension theory (Cai, 2003). The theoretical framework of Extenics is established as shown in Fig. 4.1 (Yang & Cai, 2013).

(1) Basic-element theory

The concept of basic-element integrates quality and quantity, action and relation into a triple. It is used to describe the matter, affair and relation as well as information, knowledge, problem, and strategy in a formalized way. In basic-element theory, the extensible analysis theory, conjugate analysis theory and extension transformation theory are established according to the extensibility of basic-elements.

(2) Extension set and dependent function

Extension set is to describe the change of the nature of matters, which comes into being after Cantor set and fuzzy set (Cai, 1990). It integrates the ideas of contradictory transformation and conversion in dialectics into set theory. Basic-element extension set takes both quality and quantity into consideration, providing theory foundation for addressing contradictory problems.

Dependent function, quantified calculation formula and dependent function value which represent the degree of a matter's certain nature are studied in order to analyze the changes of the nature of a matter in a quantitative way and illustrate quantitative changes and qualitative changes.

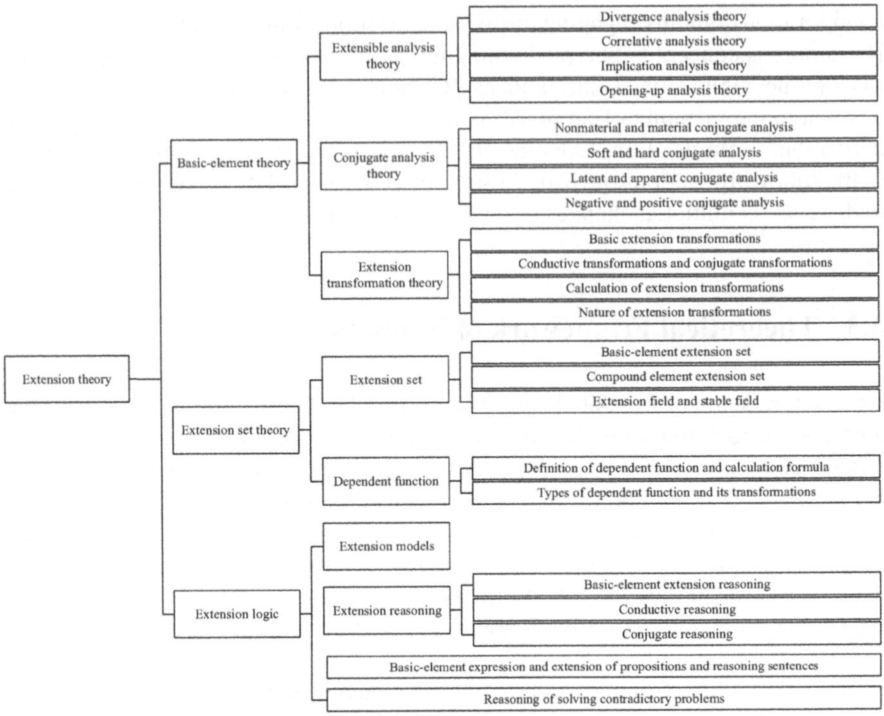

Fig. 4.1 Theoretical framework of Extenics

(3) Extension logic

The tools of solving contradictory problems are transformation and reasoning. In Extenics, the extension logic is explored that takes advantage of formalization of formal logic and the dialectical logic to study strengths of matters and their changes and becomes a kind of logic with the core of solving the transformation and reasoning of contradictory problems.

Extenics is established particularly for addressing contradiction problems. It means that specific goals cannot be reached under current conditions, which is also an obvious nature of wicked design problems. Extenics has its solid foundation set on formalizing the descriptions of matter, information, knowledge, and their relations with the real-world they engage in. Basic-elements are generated based on linguistics with a series of certain characteristics. Partial and incomplete information can be gathered and well arranged in the basic-element. With the well-structured basic-element models, the mechanism of correlation and interaction among matters can be analyzed to generate logical cause-and-effect relationships. Furthermore, extensibility and transformation rules of basic-element provide multiple methods for qualified guesses to solve the design problems. Thus, Extencis is a suitable methodology of abductive reasoning addressing the wicked design problems.

4.4 Basic-Element Theory

4.4.1 Concepts of Basic-Elements

The ordered triple in Eq. (4.1) is the mathematic model of a basic-element with O being an object, C being a characteristic, and V being the value of O about C. C and V represent the n-dimensional array, respectively (Yang & Cai, 2013).

$$B = \left(O, C, V\right) = \begin{bmatrix} Object, & c_1, & v_1 \\ & c_2, & v_2 \\ & \vdots & \vdots \\ & c_n, & v_n \end{bmatrix} \tag{4.1}$$

where

$$C = \begin{bmatrix} c_1 \\ c_2 \\ \vdots \\ c_n \end{bmatrix}, V = \begin{bmatrix} v_1 \\ v_2 \\ \vdots \\ v_n \end{bmatrix}.$$

If the basic-element is function of parameter t (t could be time, space, or other parameters), it is denoted as:

$$B(t) = \left(O(t), C, V(t)\right) = \begin{bmatrix} Object(t), & c_1, & v_1(t) \\ & c_2, & v_2(t) \\ & \vdots & \vdots \\ & c_n, & v_n(t) \end{bmatrix} \tag{4.2}$$

Basic-element is divided into matter-element, affair-element, and relation-element according to the type of the object that represented.

The matter-element describes an object (as a noun) with an n-dimensional array comprise of the matter O_m, the n-characteristics $c_{m1}, c_{m2}, \ldots, c_{mn}$, and the corresponding measures $v_{m1}, v_{m2}, \ldots, v_{mn}$, as seen in Eq. (4.3).

$$M = \left(O_m, C_m, V_m\right) = \begin{bmatrix} O_m, & c_{m1}, & v_{m1} \\ & c_{m2}, & v_{m2} \\ & \vdots & \vdots \\ & c_{mn}, & v_{mn} \end{bmatrix} \tag{4.3}$$

Interaction between matters is referred to as affair, described by affair-element. The ordered triple composed of action O_a, action's characteristics $c_{a1}, c_{a2}, \ldots, c_{an}$, and the corresponding measures $v_{a1}, v_{a2}, \ldots, v_{an}$. Action usually comes as a verb-noun pair. Basic characteristics of the action include dominating object, acting object, receiving object, time, location, degree, mode, and tool, etc. Equation (4.4) is the general form of an affair-element.

$$A = \begin{bmatrix} O_a, & c_{a1}, & v_{a1} \\ & c_{a2}, & v_{a2} \\ & \vdots & \vdots \\ & c_{an}, & v_{an} \end{bmatrix} = \begin{bmatrix} O_a, & dominating\ object & v_{a1} \\ & acting\ object, & v_{a2} \\ & receiving\ object, & v_{a3} \\ & time, & v_{a4} \\ & location, & v_{a5} \\ & degree, & v_{a6} \\ & mode, & v_{a7} \\ & tool, & v_{a8} \\ & \vdots & \vdots \end{bmatrix} \tag{4.4}$$

Relation-element is a formalized tool for describing relations. Composed of the relation name O_r, the n-characteristics $c_{r1}, c_{r2}, \ldots, c_{rn}$, and the corresponding values $v_{r1}, v_{r2}, \ldots, v_{rn}$, the n-dimensional array given in Eq. (4.5) defines a relation-element. The characteristics usually include antecedent, consequent, degree, and maintaining mode according to the basic syntax analysis. Antecedent and consequent represent the objects between which the relationship occurs, i.e., the relation-element is to describe the relation between the antecedent and consequent. Degree describes how serious the impact is. Maintaining mode describes in which way the relationship maintains.

$$R = \begin{bmatrix} O_r, & c_{r1}, & v_{r1} \\ & c_{r2}, & v_{r2} \\ & c_{r3}, & v_{r3} \\ & c_{r4}, & v_{r4} \\ & \vdots & \vdots \end{bmatrix} = \begin{bmatrix} coupling, & antecdent, & v_{r1} \\ & consequent, & v_{r2} \\ & degree, & v_{r3} \\ & maintaining\ mode, & v_{r4} \\ & \vdots & \vdots \end{bmatrix} \tag{4.5}$$

4.4.2 Logical Operations of Basic-Elements

(1) AND operation

Given two basic-elements B_1 and B_2, And operation of B_1 and B_2 refers to achieving B_1 and B_2 at the same time, denoted as $B_1 \wedge B_2$.

(2) OR operation

Given two basic-elements B_1 and B_2, OR operation of B_1 and B_2 refers to achieving at least one of B_1 and B_2, denoted as $B_1 \vee B_2$.

(3) NOT operation

Given a basic-element $B = \left(O,\ c,\ v(t) \right), v(t) \in V_0$, denoted as $B = \left(O,\ c,\ V_0 \right)$. If there is another basic-element $B_1 = \left(O,\ c,\ u \right)$, and $u \notin V_0$, it is considered that B_1 is the non-basic-element of B, denoted as $\overline{B} = B_1$. The operation of transforming B to $\overline{B}$ is referred to as NOT operation, denoted as $\neg B$, as:

$$\neg B = \left\{ \overline{B} \right\} = \left\{ \overline{B} | \overline{B} = \left(O,\ c,\ u \right), u \notin V_0 \right\} \tag{4.6}$$

4.5 Features of Basic-Elements

The basic-element has the features of correlation, implication, divergence, opening-up and conjugation. The features reveal relationships between the characteristics of the basic-element and relationships between different basic-elements. Analysis about the features facilitate the innovative design thinking with multiple creative thinking directions.

(1) Feature of correlation

In Extenics, the correlation is defined as: "The certain dependence between the measures of one basic-element and another basic-element about a certain evaluated characteristic, or between the measures of the same basic-element or of the basic-element of the same group about certain evaluated characteristics, if any, is referred to as correlation."

Given two basic-element sets, denoted as $\{B_1\}$ and $\{B_2\}$, for any basic-element $B_1 \in \{B_1\}$, if there exist at least one $B_2 \in \{B_2\}$ that makes $c_0(B_2) = f(c_0(B_1))$, $\{B_1\}$ and $\{B_2\}$ are correlative about the evaluated characteristic c_0, denoted as.

$$\{B_1\} \overset{\sim}{\rightarrow} (c_0)\{B_2\}$$

Note that the "$f(\cdot)$" herein is not limited to mathematic function, and it is used to represent that there exist some dependence relationship between the two parameters in the equation.

If $c_0(B_2) = f(c_0(B_1))$, and $c_0(B_1) = f^{-1}(c_0(B_2))$, then $\{B_1\}$ and $\{B_2\}$ are mutually correlative about the evaluated characteristic c_0, denoted as $\{B_1\} \sim (c_0)\{B_2\}$. Particularly, correlation is represented with the symbol "$\sim$" generally in application, and the symbol "$\overset{\sim}{\rightarrow}$" is only used when directivity should be indicated specially.

Generally, the sets $\{B_1\}$ and $\{B_2\}$ contain only one element and the correlation between the two basic-elements is denoted as $B_1 \sim (c_0)B_2$, or $B_1 \overset{c}{\sim} B_2$ for short.

Given two evaluated characteristics c_{01} and c_{02} about a basic-element B_1, if $c_{01}(B_1) = f(c_{02}(B_1))$, and $c_{02}(B_1) = f^{-1}(c_{01}(B_1))$, the characteristics c_{01} and c_{02} are defined to be mutually correlative about B_1.

If $c_{01}(B_1) = f(c_{02}(B_2))$ and $c_{02}(B_2) = f^{-1}(c_{01}(B_1))$, the basic-elements B_1 and B_2 are defined to be mutually correlative about the evaluated characteristics c_{01} and c_{02}.

Example 4.1 In a device with two gears engaged, the two gears are expressed with basic-elements as B_1 and B_2:

$$B_1 = \begin{pmatrix} O_1, & c_1, & v_1 \\ & c_2, & v_2 \end{pmatrix} = \begin{pmatrix} gear\ 1, & pitch, & v_1 \\ & module, & v_2 \end{pmatrix} \tag{4.7}$$

$$B_2 = \begin{pmatrix} O_2, & c_1', & v_1' \\ & c_2', & v_2' \end{pmatrix} = \begin{pmatrix} gear2, & pitch, & v_1' \\ & number\ of\ teeth, & v_2' \end{pmatrix} \tag{4.8}$$

Firstly, the pitches of gear 1 and gear 2 should be equal to ensure the engagement, i.e., B_1 and B_2 are mutually correlative about the characteristic "pitch" (c_1 and c_1' are the same), denoted for brief as:

$$B_1 \overset{c_1}{\sim} B_2 \tag{4.9}$$

Moreover, the module (m) of the gear is correlated to the pitch (p) of it (m = p/π), i.e., the characteristics c_1 (pitch) and c_2 (module) in B_1 are mutually correlative:

$$c_1(B_1) \sim c_2(B_1) \tag{4.10}$$

On the other hand, the module of gear 1 are also related to the teeth number of gear 2 considering the engagement, that is, B_1 and B_2 are also mutually correlative about the evaluated characteristics c_2 (module) and c_2' (number of teeth), denoted as:

$$c_2(B_1) \sim c_2'(B_2) \tag{4.11}$$

(2) Feature of implication

Suppose B_1 and B_2 are two basic-elements. If B_1 is realized with the realization of B_2, then B_1 implies B_2, denoted as $B_1 \Rightarrow B_2$. B_1 is called an inferior basic-element while B_2 is termed a superior basic-element. A superior basic-element can be realized with the realization of a set of inferior basic-elements. The realization of B_i is generally denoted as "B_i @".

If B_1 @ exists inevitably with the existence of B_2 @ under some certain condition l, it is referred to as B_1 implies B_2, denoted as $B_1 \Rightarrow (l)B_2$.

If both B_1 and B_2 are realized inevitably with the realization of B, it is referred to as AND Implication of B by B_1 and B_2, denoted as

$$B_1 \wedge B_2 \Rightarrow B \tag{4.12}$$

If either B_1 or B_2 is realized inevitably with the realization of B, it is referred to as OR Implication of B by B_1 and B_2, denoted as

$$B_1 \vee B_2 \Rightarrow B \tag{4.13}$$

If B is realized inevitably with the realization of both B_1 and B_2, it is referred to as AND Implication of B_1 and B_2 by B, denoted as

$$B \Rightarrow B_1 \wedge B_2 \tag{4.14}$$

If B is realized inevitably with the realization of either B_1 or B_2, it is referred to as OR Implication of B by B_1 and B_2, denoted as

$$B \Rightarrow B_1 \vee B_2 \tag{4.15}$$

The definitions can be expanded to general forms as:

$$B \Rightarrow \bigwedge_{i=1}^{n} B_i,\ B \Rightarrow \bigvee_{i=1}^{n} B_i,\ \bigwedge_{i=1}^{n} B_i \Rightarrow B,\ \bigvee_{i=1}^{n} B_i \Rightarrow B$$

If $B_1 \Rightarrow B_2$, $B_2 \Rightarrow B_3$, it implies that $B_1 \Rightarrow B_3$, or denoted as $B_1 \Rightarrow B_2 \Rightarrow B_3$.

If $B_{11} \wedge B_{12} \Rightarrow B_1$, $B_{21} \wedge B_{22} \Rightarrow B_2$, and $B_1 \wedge B_2 \Rightarrow B$, then $B_{11} \wedge B_{12} \wedge B_{21} \wedge B_{22} \Rightarrow B$.

If $B_{11} \vee B_{12} \Rightarrow B_1$, $B_{21} \vee B_{22} \Rightarrow B_2$, and $B_1 \vee B_2 \Rightarrow B$, then $B_{11} \vee B_{12} \vee B_{21} \vee B_{22} \Rightarrow B$.

(3) Feature of divergence

According to the divergence of basic-element, multiple basic-elements can emerge by diverging from a primary basic-element. The object (O), characteristic (c) and value (v) in the basic-element all have the feature of divergence. There are six kinds of divergence correspondingly.

(a) Divergence of "one object, multiple characteristics, and multiple values"

From a primary basic-element, multiple basic-elements with the same object and different characteristics and values can be extended, denoted as:

$$\begin{aligned} B &= (O,\ c,\ v) \\ &-\!\!\mid\{(O,\ c_1,\ v_1),(O,\ c_2,\ v_2),\ldots,(O,\ c_n,\ v_n)\} \\ &= \{(O,\ c_i,\ v_i), i = 1, 2, \ldots, n\} \end{aligned} \tag{4.16}$$

(b) Divergence of "multiple objects, one characteristic, and multiple values"

From a primary basic-element, multiple basic-elements with the same characteristic and different objects and values can be extended, denoted as:

$$\begin{aligned} B &= (O,\ c,\ v) \\ &-\!\!\mid\{(O_1,\ c,\ v_1),(O_2,\ c,\ v_2),\cdots,(O_n,\ c,\ v_n)\} \\ &= \{(O_i,\ c,\ v_i), i = 1, 2, \cdots, n\} \end{aligned} \tag{4.17}$$

(c) Divergence of "one object, multiple characteristics, and one value"

From a primary basic-element, multiple basic-elements with the same object and value and different characteristics can be extended, denoted as:

$$\begin{aligned} B &= (O,\ c,\ v) \\ &-\!\!\mid\{(O,\ c_1,\ v), (O,\ c_2,\ v), \cdots, (O,\ c_n,\ v)\} \\ &= \{(O,\ c_i,\ v), i = 1, 2, \cdots, n\} \end{aligned} \tag{4.18}$$

(d) Divergence of "multiple objects, multiple characteristics, and one value"

From a primary basic-element, multiple basic-elements with the same value and different objects and characteristics can be extended, denoted as:

$$\begin{aligned} B &= (O,\ c,\ v) \\ &-\!\!\mid\{(O_1,\ c_1,\ v), (O_2,\ c_2,\ v), \cdots, (O_n,\ c_n,\ v)\} \\ &= \{(O_i,\ c_i,\ v), i = 1, 2, \cdots, n\} \end{aligned} \tag{4.19}$$

(e) Divergence of "multiple objects, one characteristic, and one value"

From a primary basic-element, multiple basic-elements with the same characteristic and value and different objects can be extended, denoted as:

$$\begin{aligned} B &= (O,\ c,\ v) \\ &-\!\!\mid\{(O_1,\ c,\ v), (O_2,\ c,\ v), \cdots, (O_n,\ c,\ v)\} \\ &= \{(O_i,\ c,\ v), i = 1, 2, \cdots, n\} \end{aligned} \tag{4.20}$$

(f) Divergence of "one object, one characteristic, and multiple values"

From a primary basic-element, multiple basic-elements with the same object and characteristic and different values can be extended, denoted as:

$$\begin{aligned} B &= (O,\ c,\ v) \\ &-\!\!\mid\{(O,\ c,\ v_1), (O,\ c,\ v_2), \cdots, (O,\ c,\ v_n)\} \\ &= \{(O,\ c,\ v_i), i = 1, 2, \cdots, n\} \end{aligned} \tag{4.21}$$

(4) Feature of opening-up

The possibilities of composing, decomposing, and expanding/contracting affair, matter and relation are referred to respectively as composability, decomposability, and expandability/contractability, collectively, expandability.

According to composability, one basic-element can combine with another to generate new basic-element. Given a basic-element $B_1 = (O_1, c_1, v_1)$, there is at least one basic-element $B_2 = (O_2, c_2, v_2)$ to allow B_1 and B_2 to be composed into a new basic-element B, as shown in (4.22). B_2 is referred to as a composable basic-element of B_1.

$$B = B_1 \oplus B_2 = \begin{cases} (O_1, c_1 \oplus c_2, v_1 \oplus v_2) = \begin{bmatrix} O_1, & c_1, & v_1 \\ & c_2, & v_2 \end{bmatrix}, O_1 = O_2, c_1 \neq c_2 \\ (O_1 \oplus O_2, c_1, v_1 \oplus v_2), \ O_1 \neq O_2, c_1 = c_2 \\ \begin{bmatrix} O_1 \oplus O_2, & c_1, & v_1 \oplus c_1(O_2) \\ & c_2, & c_2(O_1) \oplus v_2 \end{bmatrix}, \ O_1 \neq O_2, c_1 \neq c_2 \end{cases} \tag{4.22}$$

According to decomposability, one basic-element can be decomposed to several new basic-elements that differ from the primary one. The decomposition of a basic-element $B = (O, c, c(O))$ about the characteristic c under some certain condition l is expressed by

$$(O, c, c(O))//(l)\{(O_1, c, c(O_1)), (O_2, c, c(O_2)), \cdots, (O_m, c, c(O_m))\}$$

denoted as $B//\{B_1, B_2, \cdots, B_m\}$.

According to the feature of expandability/ contractability, one basic-element can be expanded or contracted. Suppose a basic-element $B = (O, c, c(O))$, under some certain condition l, there must be a real number $\alpha(\alpha > 0)$ to allow $\alpha B = (\alpha O, c, \alpha v)$, where αO indicate the object with measure of αv. When $0 < \alpha < 1$, the basic-element B is contracted and when $\alpha > 1$, the basic-element B is expanded.

All the above transformations resulting in new basic-elements facilitate the generation of multiple possible solutions to contradictory problems.

(5) Feature of conjugation

Studies about the matter from its physical, systematic, dynamic, and antithetic properties can help people understand the matter structure and the nature of development and change of the matter more comprehensively. Correspondingly, four pairs of concepts are proposed in Extenics as: nonmaterial and material, soft and hard, latent and apparent, and negative and positive to describe the matter's constitution, referred to as matter's conjugacy.

Explorations of the matter's conjugacy and its reciprocal transformations provide multiple innovative methods for problem-solving.

4.6 Extension Innovation Methods

Extenics methodology is generated and formalized with the guide of ideological system of Extenics, which provide a series of methods to understand, analyze and solve contradictory problems from a new perspective. The Extenics methodology system is shown in Fig. 4.2.

The basic-elements as the logical cells of Extenics have the features of correlation, divergence, opening-up and conjugation, and the corresponding extensible analysis method and conjugation analysis method are derived from them. Once multiple approaches to solving a contradiction problem are emerged employing the analysis methods, the extension transformation method is then followed to formalize the process of solving the problem. The solutions along with the transformation process can be presented as an extension set. There usually could be more than one solution, and superiority evaluation method is to evaluate the solutions by considering multiple

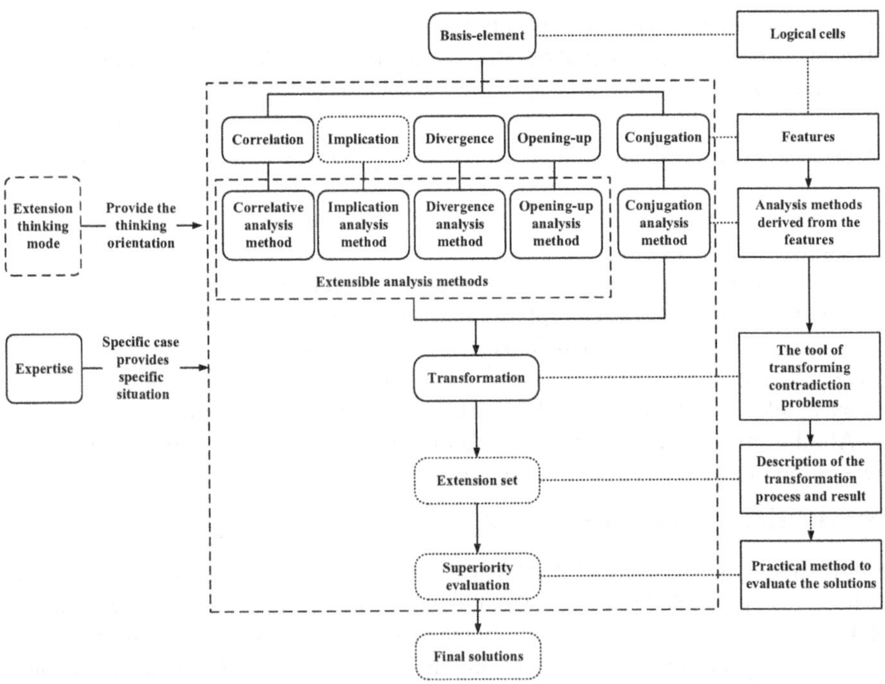

Fig. 4.2 Extenics methodology system

measure conditions to screen out the most proper solution. During this problem-solving process, the extension thinking mode provides the thinking orientation for the designer while the expertise mode provides the specific guidance with regards to the specific situation.

The followings give a brief account of the various methods found in Fig. 4.2.

(1) Extension Thinking Modes

Extension thinking mode is a specific term defined in Extenics that is a part of the Extenics innovation methods. There are four kinds of extension thinking modes as shown in Fig. 4.3. The rhombus thinking mode begins with divergence and ends with convergence. Divergence is the first step to obtain multiple solutions while convergence is to select and focus on the optimal solution among the multiple solutions. The reversed thinking mode is for thinking in the contrary way. Conjugate thinking mode is based on a matter's conjugate analysis and transformation principles. The conductive thinking mode is for solving problems indirectly with conductive transformation.

The extension thinking modes help one engage in creative thinking. They explore the laws of creative thinking with the attempt to resolve "how to innovate thinking, by considering "from where to innovate" and "how to evaluate the result of creative thinking." They define the path of thinking.

(2) Extensible Analysis Methods

The extensible analysis methods investigate the extensibility of matters, affairs, or relations. They help solve a contradiction problem by transforming the goal or condition of the problem according to the extensibilities of the matters, affairs, or relations involved. The extension rules therein are effective in helping one break away from the shackle of old thinking habit and think broadly.

(3) Conjugation Analysis Methods

The conjugation analysis methods conduct formalized and qualitative analyses of the non-material and material aspects, the soft and hard aspects, the latent and apparent aspects, and the negative and positive aspects of matters, affairs, and relations. Through evaluating the conjugates and their interactive relations and corresponding transformations, multiple strategies for solving the contradiction problem can be generated. The methods provide a new perspective for addressing contradiction problems with comprehensive analysis and brilliancy.

(4) Extension Transformation Methods

The extension transformation methods are used to transform the goals or conditions of related affairs, matters or relations to find the solution. While the methods mentioned above provide the approach to solving contradiction problems, the extension transformation methods must be implemented to realize the solution to the contradiction. There five basic transformations named substitution transformation, increasing/decreasing transformation, expansion/contraction transformation,

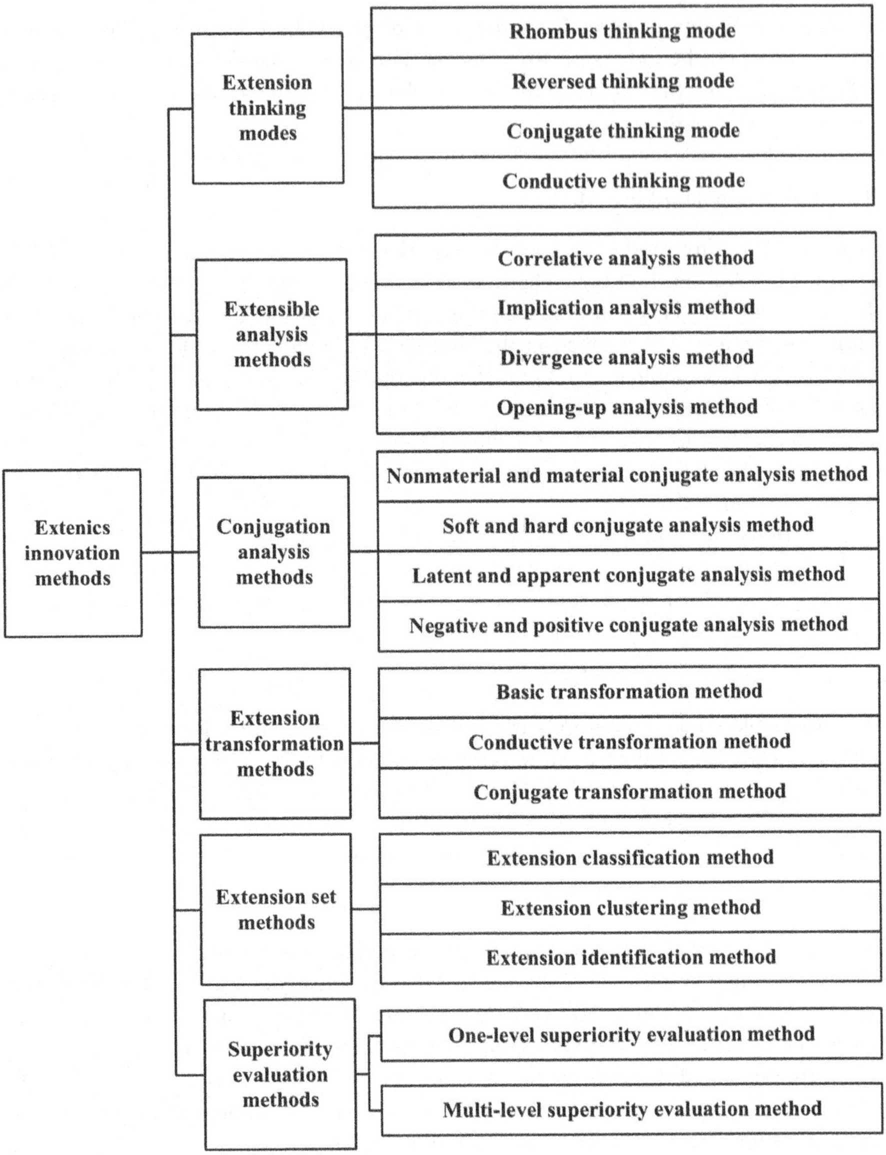

Fig. 4.3 The extension innovation method system

decomposition transformation, and duplication transformation. Transformation is an operable tool for formalizing and quantifying the process of solving contradiction problems.

(5) Extension Set Methods

The extension set methods are followed to identify and categorize objects that are dynamic and transformational. The methods describe the variability of things, using the number in between $(-\infty, +\infty)$ to describe the degree of how the thing owns certain property, and using an extensible field to describe the reciprocal transformation between the "positive" and "negative" of things.

(6) Superiority evaluation method

Superiority evaluation method is a practical method of evaluating the object, design, strategy and others by considering several measure conditions. In order to evaluate the superiority of an object, measuring indicators as well as weight coefficients should be specified first. Then the dependent degree and standard dependent degree are calculated with dependent function in Extenics. The values of dependent function can be positive or negative, so the superiority evaluation can reflect the degree of pros and cons of an object, which is practical to users.

Each of the methods has the corresponding subsets (sub-methods). Together they form the extension innovation method system. Figure 4.3 shows all the components of the extension innovation method system. The sub-methods are the specific tools for problem-solving.

The extensible analysis methods and conjugate analysis method are derived from the features of basic-elements, which are introduced in detail in the followings given their wide applicability (Cai, 1998; Yang & Cai, 2013; Cai & Yang, 2013).

(1) Extensible Analysis methods

The methods provide one with various instruments for resolving problems, including divergence analysis, correlative analysis, implication analysis, and opening-up analysis.

(a) Correlative analysis method

The correlative analysis is based on the feature of correlation for analyzing the relationships among basic-elements for a better understanding of the mechanism of correlation and interaction. It also works between the factors in the basic-elements. According to the feature of correlation, changes in one basic-element would inevitably cause changes in the other basic-element. When a contradiction problem cannot be solved upon the first attempt by a basic-element, the related ones are to be considered. The relationships between one basic-element with other basic-elements are in shape of a network structure, it is referred to as a correlation network, as shown in Fig. 4.4.

The method for solving a problem with the correlation network, named Correlation Network Method, has the following steps:

Step 1: Identify the basic-element B to be analyzed;

Step 2: List the correlation network of B according to the correlative analysis;

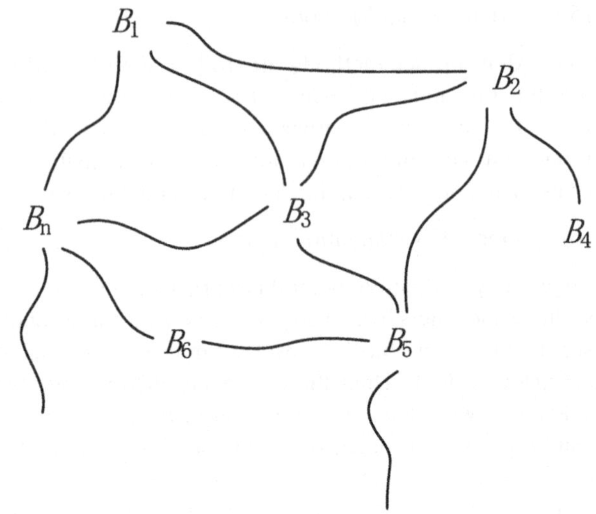

Fig. 4.4 A correlation network

Step 3: Analyze the correlation network to determine the basic-element B_i that causes the change in B, or the basic-element B_i that is changed because of the change in B;

Step 4: Apply B_i in the correlation network to solve the problem.

(b) Implication analysis method

According to the feature of implication, a basic-element can be realized with the realizations of the inferior basic-elements. The system formed using this principle is an implication system, as indicated in Fig. 4.5.

When a superior basic-element is difficult to be realized, one can look for its corresponding inferior basic-elements. When the inferior basic-elements can be easily realized, it is considered that the superior basic-element will be realized (Yang & Cai, 2013, 38). Implication analysis is apt at comprehensively studying and transforming the goal. The method for solving a problem with the implication analysis, named Implication System Method, has the following steps:

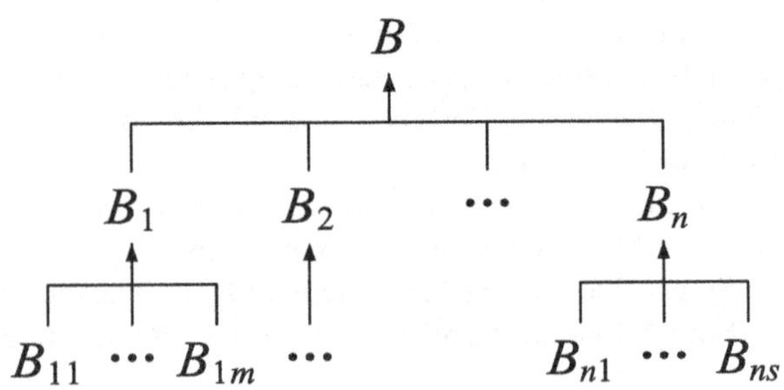

Fig. 4.5 Implication system

Fig. 4.6 Divergence tree

$$
(O,\ c,\ v) \rightarrow
\begin{cases}
\{(O_i,\ c,\ v_i), i = 1,2,3,\cdots,n\} \\
\{(O_i,\ c_i,\ v), i = 1,2,3,\cdots,n\} \\
\{(O_i,\ c,\ v), i = 1,2,3,\cdots,n\} \\
\{(O,\ c_i,\ v_i), i = 1,2,3,\cdots,n\} \\
\{(O,\ c_i,\ v), i = 1,2,3,\cdots,n\} \\
\{(O,\ c,\ v_i), i = 1,2,3,\cdots,n\}
\end{cases}
$$

Step 1: List the basic-element to be analyzed;

Step 2: Construct an implication system based on known information and the implication analysis principle;

Step 3: Based on the new information generated during the solution process, either increase or interrupt the implication system on certain layers of the system. In the case of no new information available, proceed to the next step;

Step 4: Realize the antecedent basic-element by realizing the last consequent basic-elements.

(c) Divergence analysis method

Divergence analysis is obtained from the divergence of the object (O), characteristic (c) and value (v) in a basic-element as discussed in Sect. 4.5. Per the principle of divergent analysis, multiple basic-elements can be extended from a basic-element. The diverging process forms a branching structure called a divergence tree. Figure 4.6 shows the six kinds of divergence in the divergence tree structure.

When a problem cannot be solved by using a specific basic-element, basic-elements of different objects (O), characteristics (c) or values (v) can be considered. The method for solving a problem with the divergence analysis, named Divergence Tree Method, has the following steps:

Step 1: Define goal (or condition) basic-element B to be analyzed;

Step 2: Apply proper divergence analysis method in Fig. 4.6 based on the given condition;

Step 3: Extend multiple basic-elements B_1, B_2, …, B_n from B;

Step 4: Determine the route to solve problem with extended basic-elements.

(d) Opening-up analysis method

The opening-up analysis makes use of the possibilities of composing, decomposing and expanding/contracting of the basic-element to find solutions. The design process can benefit from the broader ideas generated through the opening-up analysis. The Decomposition-combination Chain Method has following steps:

Step 1: Represent the goal and condition of the contradiction problem with the basic-element B;

Step 2: Use the Divergence Tree Method to find the combinable basic-element B_i of B or decompose B into several basic-elements B_i, $i = 1, 2,\ldots, n$;

Step 3: Investigate whether the combined or decomposed basic-elements can be used to solve the problem. If not, proceed to the next step;

Step 4: Conduct the divergence analysis, implication analysis or correlation analysis on B to obtain a set of basic-elements B_j, $j = 1, 2,\ldots, m$. In case this group of basic-elements still cannot solve the problem, consider finding their combinable basic-elements or if they can be decomposed;

Step 5: Investigate whether the combined or decomposed basic-elements can be used to solve the problem.

(2) Conjugate analysis method

Recognizing a matter from its physical, dynamic and antithetic properties can help one understand its inherent construct and discover its underlying theme of development and variation. Relevant conjugate pair methods are derived from the properties, as shown in Fig. 4.7.

All matters have a conjugate and the sum of each pair of the conjugate with its intermediate portion equals to the original matter, as is indicated in Eq. (4.23).

$$\begin{aligned} O_m &= \mathrm{re}(O_m) \oplus \mathrm{im}(O_m) \oplus \mathrm{mid}_{\mathrm{re\text{-}im}}(O_m) \\ &= \mathrm{hr}(O_m) \oplus \mathrm{sf}(O_m) \oplus \mathrm{mid}_{\mathrm{sf\text{-}hr}}(O_m) \\ &= \mathrm{lt}(O_m) \oplus \mathrm{ap}(O_m) \oplus \mathrm{mid}_{\mathrm{lt\text{-}ap}}(O_m) \\ &= \mathrm{ng_c}(O_m) \oplus \mathrm{ps_c}(O_m) \oplus \mathrm{mid}_{\mathrm{ng\text{-}ps}}(O_m) \end{aligned} \tag{4.23}$$

where

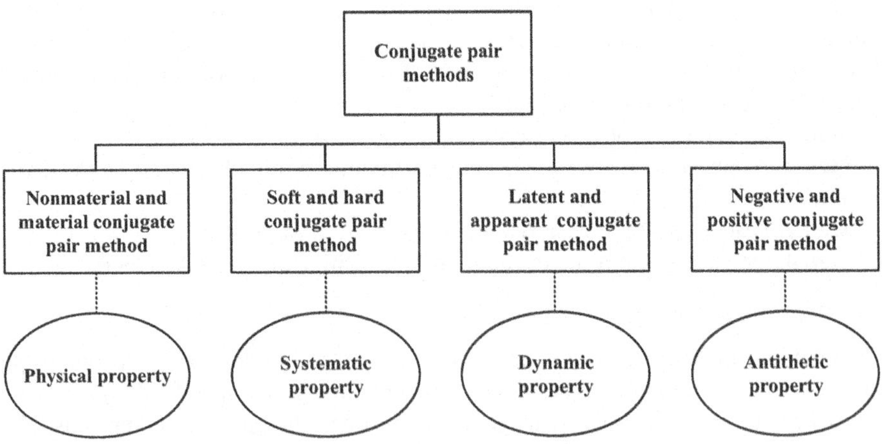

Fig. 4.7 Conjugate pair methods

O_m: the original matter;
re(O_m): the material part;
im(O_m): the nonmaterial part;
$\text{mid}_{\text{re-im}}(O_m)$: the nonmaterial-material intermediate part;
sf(O_m): soft part;
hr(O_m): hard part;
$\text{mid}_{\text{sf-hr}}(O_m)$: soft-hard intermediate part;
lt(O_m): latent part;
ap(O_m): apparent part;
$\text{mid}_{\text{lt-ap}}(O_m)$: latent-apparent intermediate part;
$\text{ng}_{\text{c}}(O_m)$: negative part;
$\text{ps}_{\text{c}}(O_m)$: positive part;
$\text{mid}_{\text{ng-ps}}(O_m)$: negative–positive intermediate part.

The material part represents things that are made up with matter and can be seen or felt. For example, the structure of a product is material part, while the function of it is nonmaterial part.

In terms of the matter's systematic property, all the components of the matter are referred to as the hard part, and the relationships between the components as well as between the matter and others are referred to as the soft part. For example, the engine and the transmission are the hard part and the relations between them are referred to as the soft part.

Considering the matter's dynamic property, any matter is changing constantly. The aspects showing up and easy to see at present are referred to as apparent part while the existing but not yet noticeable aspects are referred to as latent part. For example, the battery of an electric vehicle is an apparent part, while there exists latent risk that the battery may explode in some conditions.

Considering the matter's antithetic, the measure of the matter about the certain characteristic is a combination of the positive values and negative values. In Extenics, the part with positive values in the measurement of the matter about the certain characteristic is referred to as the positive part about the characteristic, while the part with negative values is referred to as the negative part. For example, the vehicle is applied to transport people or goods conveniently, showing the positive part to people. On the other hand, the exhaust from vehicle is harmful to human body, making it a negative part.

There may exist mid-parts between each conjugate pair that represent the critical states, respectively referred to as the nonmaterial-material intermediate part, soft-hard intermediate part, latent-apparent intermediate part, and negative–positive intermediate part.

The method for each conjugate pair has the similar steps:

Step 1: If the object to be analyzed is a matter, first express it in the form of matter-element;

Step 2: Determine the type of characteristics in the matter-element (such as nonmaterial or material characteristics);

Step 3: Conduct the corresponding conjugate pair analysis on the object and list the parts (such as nonmaterial or material part);

Step 4: Implement extension transformation of the conjugate;

Step 5: Determine the influence of the transformation on the realization of the goal. Investigate whether the problem can be solved. If not, continue the transformation till the solution is found.

The correspondence table in Table 4.1 shows the condition when a feature is to be considered and the goal each corresponding method would lead to.

4.7 Feasibility Problem in Extenics

The feasibility problem in Extenics (Cai, 1999) is defined as searching for the proper value of the given object (O_0) that belongs to a certain range (V_0), i.e.$c_0(O_0) \in V_0$. For example, when O_0 represents a chair, c_0 means the height of O_0, and the domain V_0 is defined as 'less than 50 cm', the feasibility problem is to search for a chair which height is less than 50 cm (Cai, 1994). Feasibility problem solving method articulates the design scope and boundary, object and target, which is essential to address design problem with wicked property.

The feasibility problem is represented and solved through implication function. Define the feasibility problem as $M_0 = (O_0, c_0, v_0)$. And the implication function is formulated as $M_0 \Leftarrow M_x$. It means that if M_x is realized then M_0 will be realized inevitably. Consequently, the feasibility problem is transformed to searching for M_x according to the implication function. The solution process of implication function is divided into several steps (Cai, 1994).

Step 1: Search for the set of characteristics $\{c\}$ that has determinable mutual correlations with c_0.

Step 2: Set the correlation between c and c_0 with the formulation as

$$\{f\} = \{f|c(O) = f(c_0(O)), c \in \{c\}\} \tag{4.24}$$

where 'f (·)' represents the relationships between the values of the object about c and c_0 ($c(O)$ and $c_0(O)$).

Step 3: Define the set of corresponding objects $\{O^c\}$related to O_0 about c ($c \in \{c\}$), denoted by

$$\{O^c\} = \left\{O^c \middle| O^c \overset{c}{\sim} O_0, c \in \{c\}\right\}, \ c(O^c) = v \in V \tag{4.25}$$

where the symbol '$\overset{c}{\sim}$' denotes that O^c and O_0 are correlative about c.

Step 4: Formulate the correlation between O^c and O_0 as:

$$c(O^c) = g[c(O_0)] \tag{4.26}$$

Table 4.1 Correspondence of feature with condition and goal

	Correlation	Implication	Divergence	Opening-up	Conjugation
Condition	There exist complicated relations between basic-elements	A basic-element can be achieved with the achievements of other basic-elements	The present basic-elements cannot solve the problem	Adopting merely one basic-element cannot solve the problem even with the other analysis methods	The object to be analyzed is a matter
Goal	Make clear the relations and interactions between basic-elements	Divide the target into several feasible sub-targets	Exploit more information of the present basic-elements	Develop new basic-elements based on original ones to find a way for problem solving	Help people understand the matter' structure and discover the nature of development of the matter
Method	Correlation network method	Implication system method	Divergence tree method	Decomposition-combination chain method	Conjugate pair method

which can be rewritten as Eq. (4.27).

$$v = c\left(O^c\right) = g[c(O_0)] = g[f(c_0(O_0))] = g[f(v_0)] \tag{4.27}$$

The value range of O^c, i.e. V, can be represented as:

$$V = \{v|v = g[f(v_0)], v_0 \in V_0\} \tag{4.28}$$

Step 5: Confirm the existence of M with the definition

$$M = \left(O^c, c, c(O^c)\right)c(O^c) = v \in V \tag{4.29}$$

Then the expression of O_0 can be obtained as

$$c(O_0) = g^{-1}\left[c\left(O^c\right)\right] \tag{4.30}$$

$$c_0(O_0) = f^{-1}[c(O_0)] = f^{-1}g^{-1}\left[c\left(O^c\right)\right], c_0(O_0) \in V_0 \tag{4.31}$$

'f (·)' represents the relationship between $c(O)$ and $c_0(O)$), consequently, 'f^{-1}(·)' represents the inverse relationship between $c(O)$ and $c_0(O)$. Similarly, 'g (·)' represents the relationship between $c(O^c)$ and $c(O_0)$, while 'g^{-1}(·)' represents the inverse relationships between them. Herein, the correlations between O^c and O_0, and between c and c_0 are mutual, making the inverse functions 'f^{-1}(·)' and 'g (·)' exist. Equation (4.31) means that the existence of M can lead to the result that the value of O_0 ($v_0 = c_0(O_0)$) belong to V_0, i.e. the realization of M_0. Consequently, the solution set of the implication function '$M_0 \Leftarrow M_x$'is:

$$\{M_x\} = \left\{M|M = \left(O^c, c, c(O^c)\right), O^c \in \left\{O^c\right\}, c(O^c) = v \in V\right\}.$$

4.8 Discussion and Summary

In this chapter, wicked design problems were discussed and analyzed in detail. It was argued that that abductive reasoning is required to address wicked problems. The Extenics theory was introduced to substantiate that demonstrating the property of abductive reasoning is highly suitable for addressing wicked design problems. Wicked problems can be mitigated rather than solved by searching for better solutions, making the design solutions divergent though not necessarily unique. The ill-defined and contradicting problems as well as interconnected and incomplete information are ubiquitous during the design process. The lack of definitive formula and structured expression aggravate the degree of complexity of the problem. Abductive reasoning is to make logical explanations with partial and incomplete information, allowing designers to generate good solutions. Constructed for addressing contradictory problems, Extenics possesses and explores the essence of abductive reasoning. Available information is gathered and formed into basic-elements, which are structured models with ordered triples in which the characteristics are derived from linguistics. The extension analysis and innovation methods are applied to uncover the hidden while also inherently implicated relative information and study the interconnected relationships for their root causes and induced effects. Furthermore, the solution space is extended, providing multiple viable solutions for designers to consider as to which one satisfies the design objective and meets the design requirement the best. Extenics also provides superior evaluation method to help evaluate as to which solution is better. For all the above reasons, Extenics is considered to be viable for solving wicked design problems. Synergizing Extenics with the AD theory would result in a set of powerful innovative design thinking principles. The construction of the creative synergy and the essence of the resulted design principles are presented in the following chapters.

References

Buchanan, R. (1992). Wicked problems in design thinking. *Design Issues, 8*(2), 5–21.

Cai, W. (1990). Extension set and non-compatible problems. *Advances in applied mathematics and mechanics in China*, 1–21.

Cai, W. (1994). *Matter-element model and its application*. Science and Technology Literature Publishing House.

Cai, W. (1998). Introduction of extenics. *Systems Engineering-Theory & Practice, 18*(1), 76–84.

Cai, W. (1999). Extension theory and its application. *Chinese science bulletin, 44*(17), 1538–1548.

Cai, W., Yang, C. Y., and He, B. (2001). Several problems on the research of extenics. *Journal of Guangdong University of Technology, 18*(1), 1–5.

Cai, W., Yang, C. Y., & He, B. (2003). *Preliminary extension logic*. Science Press.

Cai, W., Yang, C. Y. and Wang, G. H. (2005). A new cross discipline-extenics. *Science Foundation in China, 1*, 55–61.

Cross, N. (2006). *Designerly ways of knowing*. Springer.

Cai, W., and Shi, Y. (2006). Extenics: its significance in science and prospects in application. *Journal of Harbin Institute of technology, 38*(7), 1079–1086.

Cai, W., and Yang, C. Y. (2013). Basic theory and methodology on extenics. *Chinese science bulletin, 58*(13), 1190–1199.

Rittel, H., & Webber, M. (1973). Dilemmas in a general theory of planning. *Policy Sciences, 5*, 155–169.

Schön, D. (1983). *The reflective practitioner: How professionals think in action*. Basic Books.

Weber, E. P., & Khademian, A. M. (2008). Wicked problems, knowledge challenges, and collaborative capacity builders in network settings. *Public Administration Review, 68*(2), 334–349.

Yang, C. Y., & Cai, W. (2013). *Extenics: Theory, method and application*. Science Press.

Chapter 5
Design Innovation by Synergy

Design thinking is characterized by reframing a design problem in a structured manner and engaging an iterative design process to identify the real need. The AD theory and Exteniscs each feature some of the characteristics but not all. For instance AD does not provide an extension method as does Extenics viable for generating innovative solution and strategy. All the characteristics can be realized by a creative synergy of the two design theories. The synergy is introduced in the present chapter along with the resulted principles of innovative design thinking.

5.1 Complementary Properties of Axiomatic Design and Extenics

(1) Similarity in the objectives of the two theories

The ultimate goal of the AD theory is to establish a scientific basis to improve design activities by providing designers with logical and rational thinking process and tools. Design activities require constant iterations between "what we want to achieve" and "what we should do to achieve the goal." It is reflected in the definition of domains and zigzag mapping process of the AD theory, where the left domain which is related to the domain on the right represents "what we want to achieve" and the right domain represents the design solution, i.e., "how we propose to satisfy the requirements specified in the left domain." Moreover, in the AD theory a designer must zigzag between domains to complete the decomposition.

In contrast, the goal of Extenics is to solve contradictory problems whose solutions are hard to come by under the current conditions. Contradictory problems are ubiquitous in various areas and difficult to be solved. The starting point of Extenics research was to establish a theory with a set of methods to help people solve contradictory problems.

W. Li et al., *Principles of Innovative Design Thinking*,
https://doi.org/10.1007/978-981-19-0485-1_5

In a sense, the process of determining "what we want to achieve" and "what we should do to achieve the goal" is also a contradictory problem for the reason that we can't get the answers to those questions based on the conditions initially given. In addition, decoupling problems defined in AD are obviously contradiction problems because the existence of coupling shows that under the current condition, the goal of "getting uncoupled design solution" cannot be achieved. Thus, there are similarities in the objectives of the AD theory and Extenics. This lays the foundation for the synergy of the two theories.

(2) Similarity in concept of implication

In AD, FRs are the Whats and DPs are the Hows (Benavides, 2011). The zigzagging for decomposing FRs and DPs forms an implication system in which the superior function FR_0 can be realized based on the realization of the inferior functions. This is illustrated in Example 5.1.

Example 5.1 In the design process of a chair, the function FR_0 of the chair can be described as "For one person to sit on". It can be satisfied based on three inferior functions FR_1, FR_2 and FR_3.

FR_1: Support a person's weight on buttock
FR_2: Support a person's back
FR_3: Keep a person above the ground

This happens to coincide with the idea of implication analysis in Extenics. The implication of basic-element in Extenics is defined as follows. Suppose B_1 and B_2 are two basic-elements while B_1 can be realized with the realization of B_2. It is described as "B_1 implies B_2" and denoted as $B_1 \Rightarrow B_2$. B_1 is called inferior basic-element and B_2 is called superior basic-element. Suppose there exist three basic-elements B, B_1 and B_2, if the realizations of both B_1 and B_2 lead to the realization of B inevitably, it is called AND implication of B by B_1 and B_2, denoted as $B_1 \wedge B_2 \Rightarrow B$ (Yang and Cai, 2013).

(3) Comparison of advantages of the two theories

The AD theory establishes the basic principles generalized from design behavior to provide designers with operation process and criteria.

The concept of domain is clearly defined in the AD theory. The design space is divided into the customer domain, functional domain, physical domain, and process domain, enabling a designer to clearly define the attributes of the problems they are facing. Moreover, the design behavior is represented by the iterative mapping decomposition between domains. It provides a set of operational procedure to guide the designer's design activities. Therefore, the AD theory is the rule for promoting the design process.

At the same time, the AD theory formulates two design axioms, i.e. the Independence Axiom and the Information Axiom, from which a series of corollaries and theorems are also derived. All of these provide the rules to judge the design scheme as well as indicate the design direction.

What is more, the concept of "coupling" is also proposed in AD. Designs are classified as coupled design, decoupled design and uncoupled design. A series of theorems provide the methods for decoupling.

Extenics has a great advantage of describing the affair, matter, and relation with the basic-element. It provides a methodology for analyzing and solving contradictory problems.

Extenics decomposes an affair (or a matter, or a relation) with a set of specific characteristics (as shown in Table 5.1) and list them in the form of a triple. The basic-element makes the description of everything in the design process formalized, standard and unambiguous.

Moreover, the methodology system of Extenics provides a set of extension innovation methods, especially suitable for the innovative problem-solving. It includes the extensible analysis methods, conjugate analysis methods, extension transformation methods, extension set methods, superiority evaluation methods, and extension thinking modes. The methodology makes the descriptions of things and problems more formalized and modeled. And the extensibility and convergence of the objects are analyzed and applied for studying the thinking process. The transformation and conductivity characteristics of the matters provide the method for transforming incompatibility to compatibility, to solve conduction of contradictory problems. Moreover, the theory reflects Chinese systematic views and thinking of wholeness with the conjugate analysis.

The design principle and design process in the AD theory are not found in Extenics, while the basic-element model and extension methodology in Extenics are also absent in the AD theory. The basic-element model can be used to describe the various objects in the design process, such as elements in the four domains, coupling problems, and so on. At the same time, the extension methodology can be used to solve the problems encountered in the design process, such as design coupling. On the other hand, the design principles and design procedure in the AD theory can be used as the basic principle of and guidance for design. In the design process, the two theories can play their respective advantage to promote design thinking and solve the design problem.

Table 5.1 Specific characteristics of each basic-element model

	Affair-element	Matter-element	Relation-element
Specific characteristics	Dominating object	Function	Antecedent
	Acting object	Property	Consequent
	Receiving object	Entity	Degree
	Time	Others	Maintaining mode
	Location		Others
	Degree		
	Mode		
	Tool		
	Others		

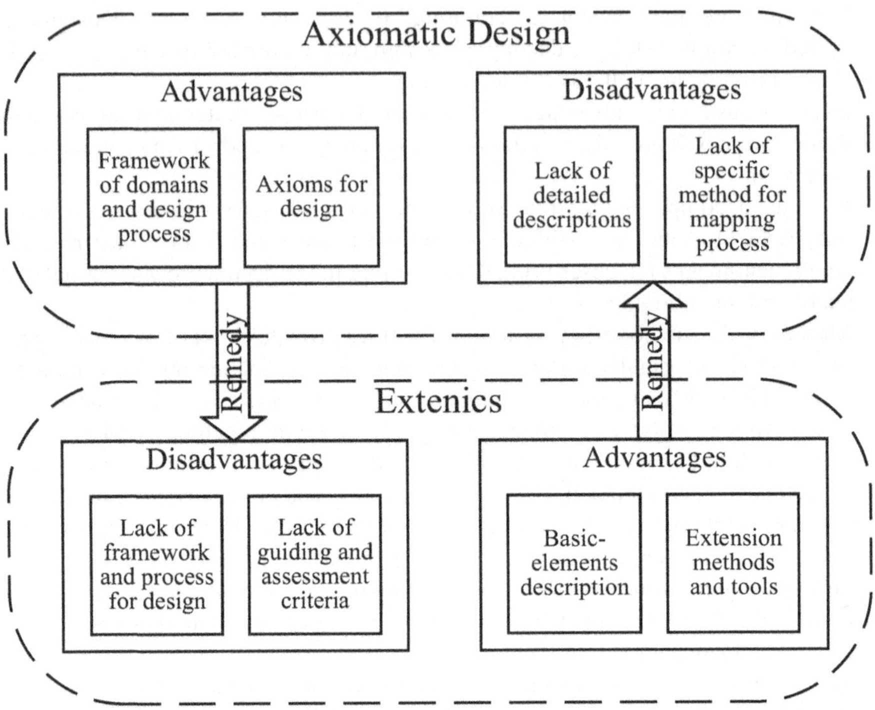

Fig. 5.1 Complementary scheme of AD and extenics

Figure 5.1 shows a schematic summarizing as to how AD and Extenics are mutually complementary. AD can be significantly enhanced by incorporating Extenics into the mapping process.

5.2 Features of Creative Synthesis

5.2.1 Functional Requirements (FRs)

A well-defined model makes it easy for designers to fill in information and determine FRs or DPs. Furthermore, the models are intended to have the same expressions while they are from different people. It is both important and difficult. Unified and unambiguous expressions can eliminate the distinctions between different designers. It makes design behaviour standard and easy to follow. There is still a long way to go to achieve this goal because of the differences in people's ideas. And a well-defined general model is the first step, which is discussed in the followings.

"Function" is a special activity or purpose of a thing or artifact. In the AD theory, the function requirement (FR) in the functional domain is a Verb-Noun pair that describes an action (Stone and Wood, 2000). In Example 5.1, the FR_0 of a chair is described as "For a person to sit on," FR_1 is "Support a person's weight on buttock," FR_2 is "Support a person's back," and FR_3 is "Keep a person above the ground." All of the FRs are in the form of Verb-Noun pairs that describe the corresponding activities.

In Extenics, affair-element is used to express the affair, which usually comes as a verb-noun pair. The affair is described from the respects of dominating object, acting object, receiving object, time, location, degree, mode, and tool.

$$A=\begin{bmatrix} O_a, & c_{a1}, & u_1 \\ & c_{a2}, & u_2 \\ & \vdots & \vdots \\ & c_{an}, & u_n \end{bmatrix}=\begin{bmatrix} O_a, & \textit{dominating object}, & u_1 \\ & \textit{acting object} & u_2 \\ & \textit{receiving object}, & u_3 \\ & \textit{time}, & u_4 \\ & \textit{location}, & u_5 \\ & \textit{degree}, & u_6 \\ & \textit{mode}, & u_7 \\ & \textit{tool}, & u_8 \\ & \vdots & \vdots \end{bmatrix}$$

Since the concept of functional requirement in the AD theory and the concept of affair in Extenics have the same form of expression, i.e., verb-noun pair, the FR can be expressed with an affair-element for the moment as

$$\mathrm{FR}=\begin{bmatrix} O_a, & \textit{dominating object}, & u_1 \\ & \textit{acting object} & u_2 \\ & \textit{receiving object}, & u_3 \\ & \textit{time}, & u_4 \\ & \textit{location}, & u_5 \\ & \textit{degree}, & u_6 \\ & \textit{mode}, & u_7 \\ & \textit{tool}, & u_8 \\ & \vdots & \vdots \end{bmatrix} \tag{5.1}$$

It's important to describe FR with a well-defined model that represent FR with a series of general and definite characteristics.

How to decide the characteristics in the affair-element model of an FR?

There are two principles for the description of the characteristics:

Principle 1: The characteristics ought to be exhaustive.

We want to establish a general model to describe the FRs even through all FRs are different. To achieve this goal, the characteristics in this general model should reflect all aspects of the FRs. An FR should be described completely after all the characteristics are considered and the model is complete. When it comes to one specific FR, some characteristics may be unimportant and can be ignored.

Principle 2: The characteristics are ought to be mutually exclusive.

It would be confusing to designers if there were overlaps between characteristics. The appropriate position for a piece of information could be difficult to determine. The appearance of a model could be influenced by subjective judgment. Thus, mutually exclusive characteristics are necessary to make the affair-element of FR practical.

The characteristics in affair-element of Extenics are abstracted from observation and research of affairs, making them representative and reliable. Furthermore, the characteristics are reviewed for FR, which is a special kind of affair.

O_a is the action of the function, which could be a verb or a verb phrase. For example, a verb could be "harvest, collect, or provide." and the verb phrases could be "convert to mechanical energy or cut into pieces."

Dominating object represents the object of the function. Acting object is the performer of the function. Receiving object is what the dominating object belongs to. The three objects are the basic characteristics of FR. For example, the function of an electric kettle, which is to raise the temperature of the water, can be expressed as:

$$\mathrm{FR} = \begin{bmatrix} \mathit{raise}, \ \mathit{dominating\ object}, & \mathit{temperature} \\ \mathit{acting\ object} & \mathit{electric\ kettle} \\ \mathit{receiving\ object}, & \mathit{water} \end{bmatrix} \tag{5.2}$$

Characteristics including time, location, degree, mode, and tool are the detailed information of the FR. They are discussed as follows.

The characteristic of time indicates the instant at which the FR performs. In practice, it is usually a relative time instead of a specific time, and it is generally in the form of descriptive phrase. For example, to describe when the FR of the cut-out device of an electric kettle performs, the time characteristic is more likely to be "after the water is boiled" rather than "10:00 am on Nov. 11."

The characteristic of location describes where the FR performs. It indicates the environment of the FR. In mechanical design, it can be a location inside a mechanical product or some specific location. For example, a transmission system is designed to transfer power from the engine to an equipment of motion. Thus, the location of the FR is "between the engine and the equipment of motion."

The degree characteristic reflects the effect of the FR. For example, a function of a ruler is to measure an object. The accuracy can be the degree of the function. In the affair-element of FR, it has the same meaning as the design range defined in the AD theory, denoted as

$$u_6(\text{FR}) = \textit{design range} \tag{5.3}$$

The mode characteristic indicates how the FR is implemented. It is a very important characteristic of FR having significant impact on the selection of corresponding DPs. Different choices of modes correspond to various kinds of design parameters. For example, a designer comes up with two different modes of driving when designing a vehicle, namely, electric drive and fuel drive. According to the two modes, the design parameter can be an electric vehicle or a gasoline-powered vehicle.

The tool characteristic in the affair-element is set to describe the thing that helps achieve the goal of the affair. It's worth noting that the products or design schemes discussed in this book are all designed by humans. They are tools and should be treated as tools applied equally to the designing of a product. A product has components and they work together to deliver the function. For example, a shovel can be called a tool that helps people lift or displace objects. Meanwhile, a bucket is not appropriate to be called a tool of excavation. Instead, it should be treated as a part of an excavator. As a result, the tool characteristic is not appropriate in the affair-element of FR. In this book the affair-element of FR is defined generally in Eq. (5.4).

$$\text{FR} = \begin{bmatrix} O_a, & \textit{dominating object}, & u_1 \\ & \textit{acting object} & u_2 \\ & \textit{receiving object}, & u_3 \\ & \textit{time}, & u_4 \\ & \textit{location}, & u_5 \\ & \textit{degree}, & u_6 \\ & \textit{mode}, & u_7 \end{bmatrix} \tag{5.4}$$

The FR_1, FR_2 and FR_3 of the chair in Example 5.1 can be expressed with affair-elements as:

$$\text{FR}_1 = \begin{bmatrix} \textit{Support}, & \textit{dominating object}, & \textit{weight} \\ & \textit{acting object} & u_2 \\ & \textit{receiving object}, & \textit{the person sit on chair} \\ & \textit{time}, & u_4 \\ & \textit{location}, & \textit{on the bottom} \\ & \textit{degree}, & u_6 \\ & \textit{mode}, & u_7 \end{bmatrix} \tag{5.5}$$

$$\text{FR}_2 = \begin{bmatrix} \textit{Support}, & \textit{dominating object}, & \textit{back} \\ & \textit{acting, object} & u_2 \\ & \textit{receiving object}, & \textit{the person sit on chair} \\ & \textit{time}, & u_4 \\ & \textit{location}, & \textit{on the back} \\ & \textit{degree}, & u_6 \\ & \textit{mode}, & u_7 \end{bmatrix} \tag{5.6}$$

$$\mathrm{FR}_3 = \begin{bmatrix} Keep, & dominating\ object, & the\ person\ sit\ on\ chair \\ & acting\ object & u_2 \\ & receiving\ object, & u_3 \\ & time, & u_4 \\ & location, & above\ the\ ground \\ & degree, & u_6 \\ & mode, & u_7 \end{bmatrix} \tag{5.7}$$

The functional requirement is elaborated by the affair-element model with the seven characteristics. This well-structured model significantly facilitates the need identification and analysis.

The non-null values in the affair-element model are generally critical parameters. So, analyzing and determining the values one by one is effective to identify the critical parameters. As discussed in Chap. 2, critical parameters often involve spatial or temporal gradients or slopes, and they are respectively corresponding to the characteristics of location and time in the affair-element model of an FR. Meanwhile, the "mode" in FR is a critical parameter that directly affects the corresponding design solution. And the "degree" in FR represents the evaluation of the FR, which is usually quantitative. It is critical that FR is investigated along with the information content.

The affair-element of FR makes it easier to reach the level of abstraction and identify the true need. The characteristics in the affair-element model provide the concrete research objects to reach the level of abstraction since all the detail information is included by the characteristics. To illustrate the concept an example of designing the engine of a car is developed as follows.

Example 5.2 A functional requirement is described as "to move a car" initially, as.

$$\mathrm{FR}_1 = \begin{bmatrix} move, & dominating\ object, & a\ car \\ & acting\ object & u_2 \\ & receiving\ object, & u_3 \\ & time, & u_4 \\ & location, & u_5 \\ & degree, & u_6 \\ & mode, & u_7 \end{bmatrix} \tag{5.8}$$

where the values from u_3 to u_7 are all null, indicating that the functional requirement statement is too vague. As discussed in Sect. 2.1.1, the abstraction of need statement should be simultaneously more general yet less vague. Thus, the values should be defined more precisely, and the functional requirement is evolved as:

$$\mathrm{FR}_1 = \begin{bmatrix} \textit{move, dominating object,} & \textit{a car} \\ \textit{acting object} & u_2 \\ \textit{receiving object,} & u_3 \\ \textit{time,} & u_4 \\ \textit{location,} & \textit{on land} \\ \textit{degree,} & \textit{with the highest speed of } 250\ \text{km/h} \\ \textit{mode,} & \textit{by burning gasoline} \end{bmatrix} \quad (5.9)$$

The abstraction process requires that the values of characteristics in the affair-element model of FR be evolved through the followings: eliminating solution-specific detail, defining the problem in solution-neutral term, transforming quantitative information into qualitative information, questioning and invalidating false constraint, and increasing the technical conciseness of the statement.

In Eq. (5.9), u_5 (on land) and u_6 (with the highest speed of 250 km/h) are solution-specific details. The u_5 is a false constraint since cars could be amphibious and u_6 is quantitative rather than qualitative. u_7 (by burning gasoline) implies that the scientific principle relevant to the particular solution is to transform chemical energy of gasoline to kinetic energy of a car, which could be more general as "to transform other form of energy to kinetic energy of a car". Finally, the functional requirement emerges as

$$\mathrm{FR}_1 = \begin{bmatrix} \textit{transform, dominating object,} & \textit{energy} \\ \textit{acting object} & u_2 \\ \textit{receiving object,} & u_3 \\ \textit{time,} & u_4 \\ \textit{location,} & \textit{any path} \\ \textit{degree,} & \begin{matrix}\textit{to kinetic energy of a car with} \\ \textit{the highest acceptable speed}\end{matrix} \\ \textit{mode,} & u_7 \end{bmatrix} \quad (5.10)$$

5.2.2 Design Parameters (DPs)

Design parameter (DP) in the AD theory is the physical variable that satisfies the functional requirement. DPs are expressed in nouns, obviously. Similarly, the concept of matter in Extenics is also expressed as a noun. The matter-element describes an object (as a noun) with an n-dimensional array comprise of the matter O_m, the n-characteristics c_{m1}, c_{m2},...,c_{mn}, and the corresponding measures v_1, v_2,..., v_n. In Extenics, a matter is usually described with function, property and entity characteristics. When concept design process is considered, the characteristics of function, principle, structure, shape and arrangement should be contemplated thoroughly to describe a product (Zhao, 2001), as shown in Eq. (5.11).

$$M = \begin{bmatrix} O_m, & c_{m1}, & v_1 \\ & c_{m2}, & v_2 \\ & \vdots & \vdots \\ & c_{mn}, & v_n \end{bmatrix} = \begin{bmatrix} O_m, & function, & v_1 \\ & principle, & v_2 \\ & structure, & v_3 \\ & shape, & v_4 \\ & arrangement & v_5 \\ & \vdots & \vdots \end{bmatrix} \tag{5.11}$$

A DP can be expressed with the matter-element model. To make the model of DP more appropriate, the characteristics of DP are redefined so as to be exhaustive and mutually exclusive, similar to the principles followed while selecting the characteristics of FR in Sect. 5.2.2.

The characteristic of function is the function of DP. According to the discussion made in Sect. 5.2.2, functional requirement (FR) can be expressed with the affair-element model. Functional requirement is an expected function while the function of DP is an actual function, thus they have the same expression of affair-element model. The function of DP can be expressed as

$$v_1(\mathrm{DP}) = \begin{bmatrix} O_a, & dominating\ bject, & u_1 \\ & acting\ object & u_2 \\ & receiving\ object, & u_3 \\ & time, & u_4 \\ & location, & u_5 \\ & degree, & u_6 \\ & mode, & u_7 \end{bmatrix} \tag{5.12}$$

Herein, the function of DP is an objective while FR is a subjective requirement. It is worth noting that they are not always the same. The difference is usually related to the degree of abstraction, design scopes, or the envelope within which the design is bound. For example, one function of a bottle is providing a closed space, while the FR could be containing water or containing sand or something else. The characteristic of degree in Eq. (5.12) reflects the effect of the function. As discussed above, in the affair-element model of FR, the degree has the same meaning as the design range in the AD theory. Similarly, the characteristic of degree in Eq. (5.12) equals to the system range in AD, denoted as

$$u_6(\mathrm{DP}) = system\,range \tag{5.13}$$

The characteristic of principle explains how DP works to achieve the corresponding function. "Principle" usually represents a general law that explains how thing works or why thing happens, while "mode" tends to be a particular way of doing something. But when we try to explain how a specific function works with both the general principle and the particular mode being considered, they are usually not sharply demarcated. For example, the mode "Converting electrical energy into mechanical energy" of a vehicle implies the general principle of "energy transformation". Considering the *principle 2,* in which characteristics ought to be mutually exclusive, discussed in Sect. 5.2.1, we eliminate the characteristic principle in the matter-element model of DP and describe how DP works with the characteristic mode in Eq. (5.12).

We use the characteristic configuration to represent DP itself, the physical parameter, which can be extended to shape, structure, size, material, etc., using the followings:

$$(O_m,\ configuration,\ v) - | \left\{ \begin{array}{c} (O_m,\ shape,\ v), \\ (O_m,\ structure,\ v), \\ (O_m,\ size,\ v), \\ (O_m,\ material,\ v), \\ \vdots \end{array} \right\} \tag{5.14}$$

Meanwhile, the characteristic arrangement is set to describe the particular position of the DP in the whole system and the relationship between the DP and other DPs, that can be extended to position, assembly mode, and others, as:

$$(O_m,\ arrangement,\ v) - | \left\{ \begin{array}{c} (O_m,\ position,\ v), \\ (O_m,\ assembly\ mode,\ v), \\ \vdots \end{array} \right\} \tag{5.15}$$

It's worth noting that the configuration characteristic contains the relationships between the elements of a DP, while the arrangement characteristic reflects the relationships between the DP and other DPs.

The matter-element model of DP is formally defined as follows

$$\mathrm{DP} = \begin{bmatrix} O_m, & function, & v_1 \\ & configuration, & v_2 \\ & arrangement, & v_3 \end{bmatrix} \tag{5.16}$$

The configuration and arrangement are basic characteristics of a DP, which determine other characteristics such as physical properties, chemical properties, cost, lifetime and so on. For example, the material of a matter determines its conductivity, hardness, ductility among others. The characteristics of cost and lifetime are usually determined by material, size, structure and so on. These kinds of characteristics are defined as "*intermediate characteristics*". They are determined by the basic characteristics, and they can determine many aspects of the function. The basic characteristics are independent variables while the intermediate characteristics are dependent variables, denoted as

$$\varphi = g(v_2, v_3) \tag{5.17}$$

where φ represents intermediate characteristics and $g(\cdot,\cdot)$ represents the relationship between the basic characteristics and intermediate characteristics.

Meanwhile, the function is also a dependent variable related to the basic characteristics and intermediate characteristics, denoted as

$$v_1 = h(v_2, v_3, \varphi) = h(v_2, v_3, g(v_2, v_3)) \tag{5.18}$$

It comes down to the fact that the function is dependent upon the basic characteristics (v_1 and v_2), which can be simplified as:

$$v_1 = f(v_2, v_3) \tag{5.19}$$

According to the AD theory, DP is the critical variable in the physical domain that characterizes the design as meeting the specified FR. From the discussion above, it is found that there are many sub-parameters of a DP (such as basic characteristics and intermediate characteristics and so on). DP must satisfy the corresponding FR, and this satisfaction is achieved through certain key sub-parameters of the DP. They can determine whether the FR could be satisfied, or they can determine the values of some aspects of the FR. They are necessary and need be well-defined to satisfy the corresponding FR. Meanwhile, there also exist certain unimportant sub-parameters of DP when satisfying the corresponding FR.

Considering sub-parameters for a DP is more in line with reality. For example, an FR is to provide elastic force and the corresponding DP is determined to be a mechanical spring. The spring has various sub-parameters including length, rotation, material, wire diameter, color, conductivity, and so on. Many of the sub-parameters are relevant to providing elastic force, such as length, rotation, material and wire diameter. Meanwhile, the color and conductivity of the spring seem to be irrelevant to providing elastic force. Therefore, it's important to select the proper sub-parameters to construct the matter-element model of DP.

Example 5.3 In Example 5.1, there are three DPs to satisfy FR_1 (Support a person's weight on buttock), FR_2 (Support a person's back), and FR_3 (Keep a person above the ground).

DP_1: The seat
DP_2: The back
DP_3: The leg

They can be expressed with matter-elements as:

$$DP_1 = \begin{bmatrix} \textit{The seat}, & \textit{function}, & \textit{Support a person's weight on buttock} \\ & \textit{configuration}, & \textit{Surface} \\ & \textit{arrangement}, & \textit{Horizontal} \end{bmatrix} \tag{5.20}$$

$$\mathrm{DP}_2 = \begin{bmatrix} \textit{The back}, & \textit{function}, & \textit{Support a person's back} \\ & \textit{configuration}, & \textit{Surface} \\ & \textit{arrangement}, & \textit{Vertical} \end{bmatrix} \quad (5.21)$$

$$\mathrm{DP}_3 = \begin{bmatrix} \textit{The leg}, & \textit{function}, & \textit{Keep a person above the ground} \\ & \textit{configuration}, & \textit{Column} \\ & \textit{arrangement}, & \textit{Vertical} \end{bmatrix} \quad (5.22)$$

In these basic-element models of DPs, the characteristics are abstract, which can be extended. For instance, the configuration of DPs can be extended, denoted as:

$$(\textit{the seat, configuration, surface}) - | \begin{Bmatrix} (\textit{the seat, shape}, v), \\ (\textit{the seat, size}, v), \\ (\textit{the seat, material}, v) \end{Bmatrix} \quad (5.23)$$

$$(\textit{the back, configuration, surface}) - | \begin{Bmatrix} (\textit{the back, shape}, v), \\ (\textit{the back, size}, v), \\ (\textit{the back, material}, v) \end{Bmatrix} \quad (5.24)$$

$$(\textit{the leg, configuration, surface}) - | \begin{Bmatrix} (\textit{the leg, shape}, v), \\ (\textit{the leg, size}, v), \\ (\textit{the leg, material}, v) \end{Bmatrix} \quad (5.25)$$

The extended characteristics are sub-parameters of the configuration of the seat, the back, and the leg. With the determinations of the parameters, the design of chair is completed. For example, a design solution of chair is shown in Fig. 5.2.

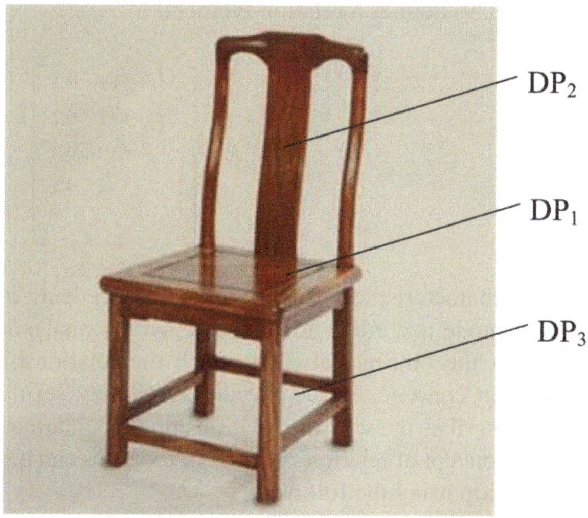

Fig. 5.2 A kind of chair

5.2.3 Coupling Problems

As discussed in Sect. 3.2, the functional requirements as defined in AD are explicitly expressed for each layer as FR = $[A]\{\mathrm{DP}\}$. Assume that there are m FRs and n DPs as a general design case, the design matrix is

$$[A] = \begin{bmatrix} A_{11} & A_{12} & \cdots & A_{1n} \\ A_{21} & A_{22} & \cdots & A_{n2} \\ \vdots & \vdots & & \vdots \\ A_{m1} & A_{m2} & \cdots & A_{mn} \end{bmatrix} \tag{5.26}$$

where $A_{ij} = \partial \mathrm{FR}_i / \partial \mathrm{DP}_j$. When the design matrix $[A]$ is diagonal, the design is called an uncoupled design, meaning that each of the FRs can be satisfied independently by one DP. In this case,

$$A_{ij} = \partial \mathrm{FR}_i / \partial \mathrm{DP}_j = 0 \ (i \neq j) \tag{5.27}$$

Conversely, when a design is not an uncoupled design, some off-diagonal elements are nonzero elements, denoted as:

$$\exists A_{ij} = \partial \mathrm{FR}_i / \partial \mathrm{DP}_j \neq 0 \ (i \neq j) \tag{5.28}$$

With A_{ij} being the contribution made by DP_j to FR_i, coupling is a relationship between DP_j and FR_i $(i \neq j)$. In Extenics, relation-element is a formalized tool to describe relationships. It is composed of relation name O_r, n-characteristics c_{r1}, $c_{r2}, \ldots, c_{rn}$, and corresponding values $w_1, w_2, \ldots, w_n$, the n-dimensional array given in Eq. (5.29) defines a relation-element.

$$R = \begin{bmatrix} O_r, & c_{r1}, & w_1 \\ & c_{r2}, & w_2 \\ & c_{r3}, & w_3 \\ & c_{r4}, & w_4 \\ & \vdots & \vdots \end{bmatrix} \tag{5.29}$$

The characteristics usually include antecedent, consequent, degree, and maintaining mode according to the basic syntax analysis. Antecedent and consequent represent the objects between which the relationship occurs. Antecedent has an impact on consequent. Degree describes how serious the impact is. Maintaining mode describes the way the relationship is maintained.

The concept of relation-element in Extenics can be applied to describe a coupling relationship using the followings

$$R = \begin{bmatrix} coupling, & antecedent, & w_1 \\ & consequent, & w_2 \\ & degree, & w_3 \\ & maintaining\ mode, & w_4 \\ & \vdots & \vdots \end{bmatrix} \tag{5.30}$$

The coupling is between DP_j and FR_i $(i \neq j)$, and A_{ij} is to describe the contribution made by DP_j to FR_i, thus employing the expression of relation-element, one has

$$w_1 = \mathrm{DP}_j \tag{5.31}$$

$$w_2 = \mathrm{FR}_i \tag{5.32}$$

$$w_3 = A_{ij} \tag{5.33}$$

The relation-element of the coupling is denoted as R below:

$$R = \begin{bmatrix} coupling, & antecedent, & \mathrm{DP}_j \\ & consequent, & \mathrm{FR}_i \\ & degree, & A_{ij} \\ & maintaining\ mode, & w_4 \\ & \vdots & \vdots \end{bmatrix} \tag{5.34}$$

Decoupling the coupled design is to alter the relationship of coupling. Then the characteristic "maintaining mode" becomes the key to decouple the coupling, which would require a detailed model in the form of basic-element. The basic-element models of FR and DP (Eqs. 5.4 and 5.16) provide abundant information for the objective.

Generally, the coupling problem occurs under certain conditions, such as a specific circumstance, a specific time or location. The conditions that underlie functional coupling are important for finding viable decoupling solutions. Moreover, for further study on the essential impact of DP_j on FR_i, it is necessary to obtain the characteristics of antecedent and consequent when a coupling occurs, as well as the specific impact. The basic-element model of the maintaining mode of coupling is shown in Eq. (5.35).

$$M = \begin{bmatrix} w_4, & coupling\ condition, & p_1 \\ & characteristic\ of\ antecedent, & p_2 \\ & characteristic\ of\ consequent, & p_3 \\ & impact, & p_4 \\ & \vdots & \vdots \end{bmatrix} \tag{5.35}$$

where p_1 is related to circumstance, time, location, and etc. And p_2 is related to the basic or intermediate characteristics of DP_j. Similarly, p_3 is related to the set of characteristics of FR_i.

Decoupling aims to eliminate the impact that DP_j has on FR_i, with the realization of Eq. (5.27). It means the degree of coupling should be reduced to zero. The objective can be expressed as

$$R' = \begin{bmatrix} coupling, & antecedent, & DP_j \\ & consequent, & FR_i \\ & degree, & 0 \\ & maintaining\ mode, & w'_4 \\ & \vdots & \vdots \end{bmatrix} \tag{5.36}$$

$$M' = \begin{bmatrix} w'_4, & coupling\ condition, & p'_1 \\ & characteristic\ of\ antecedent, & p'_2 \\ & characteristic\ of\ consequent, & p'_3 \\ & impact, & p'_4 \\ & \vdots & \vdots \end{bmatrix} \tag{5.37}$$

5.2.4 Constraints

In the design process, the most important objective is to realize the required function through the underlying design parameters. In addition, the design objective should always satisfy the constraints (Cs), which sets the design solution boundaries. In the AD theory, the constraints are divided into input constraints and system constraints. Input constraints are imposed as part of the design specifications, while system constraints are imposed by the system within which the design solution functions.

Constraints often run through the whole process of design and have an impact on the subsequent steps and design details, such as the selection of configurations and the determination of component parameters. At the same time, in the design process, new system constraints are constantly introduced, and designers need to analyze and manage them so that they are not violated. Design functions focus on the completion of action, while constraints focus on the performance requirements of the function and design scheme. Since there are no corresponding specific design parameters to restrain, they are reflected in the decomposed sub-functions and corresponding design parameters.

The affair-element for FR_i and matter-element for DP_i as established in Sects. 5.2.1 and 5.2.2 are

$$\mathrm{FR}_i = \begin{bmatrix} O_a, & \textit{dominating object}, & u_1 \\ & \textit{acting object} & u_2 \\ & \textit{receiving object}, & u_3 \\ & \textit{time}, & u_4 \\ & \textit{location}, & u_5 \\ & \textit{degree}, & u_6 \\ & \textit{mode}, & u_7 \end{bmatrix}, \quad \mathrm{DP}_i = \begin{bmatrix} O_m, & \textit{function}, & v_1 \\ & \textit{configuration}, & v_2 \\ & \textit{arrangement}, & v_3 \end{bmatrix}.$$

At each level of decomposition, constraints can be added to the corresponding basic-element model or embodied in the requirements of characteristics, making the constraints a part of FR_i and DP_i. The following example helps to illustrate this point.

Example 5.4 In the design of a power machine, an imposed constraint is that the power is no <20 KW, which directly limits the scope of engine selection. The basic-element model of engine can be expressed as follows:

$$\mathrm{FR} = \begin{bmatrix} \begin{matrix}\textit{Convert to}\\ \textit{kinetic energy},\end{matrix} & \textit{dominating object}, & \textit{chemical energy} \\ & \textit{acting object} & \textit{engine} \\ & \textit{receiving object}, & \textit{fuel} \\ & \textit{time}, & u_4 \\ & \textit{location}, & \textit{inside the engine} \\ & \textit{degree}, & \begin{matrix}\textit{the power is not}\\ < 20\ \mathrm{kw}\end{matrix} \\ & \textit{mode}, & \textit{burn} \end{bmatrix} \tag{5.38}$$

$$\mathrm{DP} = \begin{bmatrix} \textit{Diesel engine}, & \textit{function}, & \begin{matrix}\textit{Converting chemical}\\ \textit{energy into kinetic energy}\end{matrix} \\ & \textit{configuration}, & v_2 \\ & \textit{arrangement}, & v_3 \end{bmatrix} \tag{5.39}$$

The configuration characteristic v_2 and the arrangement characteristic v_3 of the diesel engine are limited by the input constraints. Suppose there are several alternatives at present for DP in configuration and arrangement,

$$v_2 = \{v_{21}, v_{22}, \cdots, v_{2n}\} \tag{5.40}$$

$$v_3 = \{v_{31}, v_{32}, \cdots, v_{3n}\} \tag{5.41}$$

Then selection must be made considering the constraint of "the power is not <20 KW".

In order to facilitate the search for design solutions, the constraints are divided into simple constraints and complex constraints according to the relationship between constraints and functional requirements. Simple constraint refers to a constraint that acts on one sub-function and eventually reflects on one design parameter; while a complex constraint is related to multiple sub-functions and is eventually reflected on multiple design parameters. In practical application, constraints usually do not act on one specific sub-function, but rather on the overall design scope. Constraints such as cost constraint, volume constraint, life constraint, etc. are an integral part of the entire design solution. They are complex constraints. When a constraint is a simple constraint, it will ultimately act on one functional requirement and the corresponding design parameter, without impacting other functional requirements and design parameters. Such a scenario is easier to meet than when complex constraints are involved. When the constraint is a complex constraint, it is likely to affect a series of design parameters during the decomposition process. Take the volume constraint as an example, the volume size of a vehicle is the sum of all the components' volume, thus having an impact on the selection of each vehicle component.

Constrains of FRs are reflected in the corresponding design parameters during the mapping process. Moreover, constraints of FRs and DPs can lead to the emergence of constrains of FRs and DPs in sub layers. In this book, constrains are integrated into the basic-element models of FRs and DPs, which is beneficial in the following respects:

(1) Abstract and general constraints are decomposed layer by layer in the function decomposition process, eventually affecting the selection of design parameters.
(2) It's convenient for designers to consider the influence of constraints in each step of the design process.
(3) Transitive relationship of constraints can be clearly shown. The object and expression of constraints are determined in the decomposition process which is conducted layer by layer.
(4) Constraints in the primitive model can be treated as the evaluation index for the convergence of design schemes, thus helping to select the most appropriate design scheme from a variety of many.

5.3 Synergized Formulation

5.3.1 Association of Functional and Physical Domains

As discussed above, a functional requirement FR_i and its corresponding design parameter DP_i are represented as:

$$\text{FR}_i = \begin{bmatrix} O_a, & \textit{dominating object}, & u_1 \\ & \textit{acting object} & u_2 \\ & \textit{receiving object}, & u_3 \\ & \textit{time}, & u_4 \\ & \textit{location}, & u_5 \\ & \textit{degree}, & u_6 \\ & \textit{mode}, & u_7 \end{bmatrix}, \quad \text{DP}_i = \begin{bmatrix} O_m, & \textit{function}, & v_1 \\ & \textit{configuration}, & v_2 \\ & \textit{arrangement}, & v_3 \end{bmatrix}.$$

When the design is an uncoupled design, the FR is satisfied by a specific DP. The relationship between a functional requirement and its design parameter is described with a design equation according to the AD theory:

$$\text{FR}_i\$\ f_a(\text{DP}_i) \tag{5.42}$$

where "f_a (·)" means the "function of" and the symbol "\$" means "satisfies" when read from right to left. Equation (5.42) means "the function of DP_i satisfies FR_i." When DP_i and FR_i quantitatively represent the physical quantities of an engineering system, f_a (•) denotes the associated laws of physics. In such way, Eq. (5.42) can be rewritten as $\text{FR}_i = f_a(\text{DP}_i)$ (Farid and Suh, 2016).

In matter-element model of DP_i, the function of DP_i is represented as v_1, thus Eq. (5.42) can be rewritten as

$$\text{FR}_i\$\ v_1(\text{DP}_i) \tag{5.43}$$

where "$v_1(\text{DP}_i)$" means "the function value (v_1) of DP_i in the expression matrix." As discussed in Eq. (5.12), $v_1(\text{DP}_i)$ can also be expressed as an affair-element model:

$$v_1(\text{DP}_i) = \begin{bmatrix} O_a, & \textit{dominating object}, & u_1 \\ & \textit{acting object} & u_2 \\ & \textit{receiving object}, & u_3 \\ & \textit{time}, & u_4 \\ & \textit{location}, & u_5 \\ & \textit{degree}, & u_6 \\ & \textit{mode}, & u_7 \end{bmatrix} \tag{5.44}$$

It has the same form of representation as FR_i. As such they can be compared regardless if they are quantities or other kinds of variables. If they have the same affair-element models, they are considered equal to each other, denoted as:

$$\text{FR}_i = v_1(\text{DP}_i) \tag{5.45}$$

Since the function of DP is objective while FR is a subjective requirement, they are not always the same. When they are not the same, an extension transformation T is conducted to transform $v_1(\mathrm{DP}_i)$ to FR_i, using the following scheme

$$\mathrm{FR}_i = T v_1(\mathrm{DP}_i) \tag{5.46}$$

$$v_1(\mathrm{DP}_i) = T^{-1}\mathrm{FR}_i \tag{5.47}$$

where T^{-1} is the inverse transformation of T, i.e., when T is a transformation from A to B, then T^{-1} is a transformation from B to A. As an example, say one functional requirement is containing water and a bottle is designed to satisfy the FR, which has the function of providing an enclosed volume. In this case, T is a transformation from providing an enclosed volume to containing water. As another example, one FR is to measure vehicle speed, and a device is designed with the function of measuring engine speed. So, the T in this case is a transformation from measuring engine speed to measuring vehicle speed, which can be calculated with following certain physical principles and formula. The core of T is the relationship between $v_1(\mathrm{DP}_i)$ and FR_i, it could be related to physical principles or formula, or the process from the general to the particular.

In the AD theory, a design parameter is the physical variable that satisfies the functional requirement. According to Extenics, an acting object characteristic is the subject that performs the action, which has the same implication with the corresponding design parameter (DP_i) of the functional requirement (FR_i). Consequently, FR_i and DP_i have the following relation

$$\mathrm{DP}_i = u_2(\mathrm{FR}_i) \tag{5.48}$$

where "u_2 (•)" means "the value (u_2) of the acting object characteristic in the expression matrix."

The relationship between FR_i and DP_i is represented by the following expression,

$$\mathrm{FR}_i = \begin{bmatrix} O_a, \ \textit{dominating object}, & u_1 \\ \textit{acting object} & \mathrm{DP}_i \\ \textit{receiving object}, & u_3 \\ \textit{time}, & u_4 \\ \textit{location}, & u_5 \\ \textit{degree}, & u_6 \\ \textit{mode}, & u_7 \end{bmatrix} \tag{5.49}$$

With DP_i been obtained, the decomposition process is to reiterate back to the functional domain to create FR_{in} at the next level that collectively satisfies the highest-level FR_i. It is required that the decomposition of FR_i is performed according to the functions of DP_i. The function of DP_i is identified in the matter-element model. In case there are more than one function of DP_i, the characteristic "function" is expressed as "$function_1$, $function_2$,..., $function_n$" with the corresponding values being "v_{11}, v_{12},...,v_{1n}". The matter-element model of DP_i with only the functional characteristics is referred to as

$$\mathrm{DP}_i = \begin{bmatrix} O_m, \ function_1, & v_{11} \\ function_2, & v_{12} \\ \vdots & \vdots \\ function_n, & v_{1n} \end{bmatrix} \tag{5.50}$$

The sub-functions of FR_i are therefore satisfied by v_{1n} (DP_i)

$$\mathrm{FR}_{in}\$v_{1n}(\mathrm{DP}_i) \tag{5.51}$$

where "v_{1n} (DP_i)" represents the n-th functional characteristic of DP_i, which also can be expressed with an affair-element model as:

$$v_{1n}\ (\mathrm{DP}_i) = \begin{bmatrix} O_a, \ dominating\ object, & u_1 \\ acting\ object & u_2 \\ receiving\ object, & u_3 \\ time, & u_4 \\ location, & u_5 \\ degree, & u_6 \\ mode, & u_7 \end{bmatrix} \tag{5.52}$$

Conversely, when FR_{in} and v_{1n} (DP_i) have the same affair-element models, they are considered equal to each other, denoted as:

$$\mathrm{FR}_{in}{=}v_{1n}(\mathrm{DP}_i) \tag{5.53}$$

In the case that they are not the same, an extension transformation T is conducted

$$\mathrm{FR}_{in} = T v_{1n}(\mathrm{DP}_i) \tag{5.54}$$

$$v_{1n}(\mathrm{DP}_i) = T^{-1}\mathrm{FR}_{in} \tag{5.55}$$

In the AD theory, after the proper DPs are chosen, one has to "zag" to the functional domain for further decomposition. The relationship between DP_i and FR_{in} during the zigzag mapping from DP_i to FR_{in} can be properly represented using the following expression,

$$\mathrm{DP}_i = \begin{bmatrix} O_m, \ function_1, & T^{-1}\mathrm{FR}_{i1} \\ function_2, & T^{-1}\mathrm{FR}_{i2} \\ \vdots & \vdots \\ function_n, & T^{-1}\mathrm{FR}_{in} \end{bmatrix} \tag{5.56}$$

The sub-functions are determined according to the DP in the higher level. Generally, the FR_{in} can be expressed with v_{1n} (DP_i) directly.

Example 5.5 One of the functional requirements (FR_0) is to display the vehicle speed to the driver, and the corresponding device (DP_0) is designed to satisfy the function. FR_0 can be expressed with the affair-element model, in which the acting object is the DP_0.

$$\mathrm{FR}_0 = \begin{bmatrix} display, \ dominating\ object, & the\ speed \\ acting\ object & \mathrm{DP}_0 \\ receiving\ object, & the\ vehicle \\ time, & at\ all\ times \\ location, & in\ the\ driver's\ cab \\ degree, & u_6 \\ mode, & u_7 \end{bmatrix} \tag{5.57}$$

The DP_0 is shown as follows.

$$\mathrm{DP}_0 = \begin{bmatrix} the\ device, \ function, & v_1 \\ configuration, & v_2 \\ arrangement, & v_3 \end{bmatrix} \tag{5.58}$$

where "$v_1(\mathrm{DP}_0)$" can be represented with an affair-element model as:

$$v_1(\mathrm{DP}_0) = \begin{bmatrix} display, \ dominating\ object, & the\ speed \\ acting\ object & \mathrm{DP}_0 \\ receiving\ object, & the\ vehicle \\ time, & at\ all\ time \\ location, & in\ the\ driver's\ cab \\ degree, & u_6 \\ mode, & u_7 \end{bmatrix} \tag{5.59}$$

Compare Eqs. (5.57) with (5.59), and it is easy to conclude that

$$\mathrm{FR}_0 = v_1(\mathrm{DP}_0) \tag{5.60}$$

Obviously, the vehicle speed is varying and the measurement in real time is needed. Moreover, the location of the function of displaying the speed is in the driver's cabin, which is away from the place where measurement is made. Thus, there must be a pathway between them to transmit the signal. Finally, the signal should be displayed to the driver with clarity. There are three sub-functions of DP_0, and the matter-element model of DP_0 with only the functional characteristics is denoted as

$$DP_0 = \begin{bmatrix} \textit{the device}, & \textit{function}_1, & \textit{measure the speed} \\ & \textit{function}_2, & \textit{transmit the signal} \\ & \textit{function}_3, & \textit{display the values} \end{bmatrix} \quad (5.61)$$

The sub functions in the next layer can then be decomposed accordingly.

5.3.2 Independence Criterion

As discussed earlier, the Independence Axiom of AD is embodied by the design equation and design matrix $[A]$. To determine the independence of FRs, one must make sure the design matrix is either diagonal or of a triangular form. In this step, determining the element A_{ij} in $[A]$ is essential. A_{ij} represents the contribution made by DP_j to FR_i; that is, $A_{ij} = \partial FR_i / \partial DP_j$.

How is the value of A_{ij} determined?

By definition, functional requirement (FR) is a verb-noun pair that describes an action while design parameter (DP) is a noun that describes an object. They are different in nature. Oftentimes there are issues associated with calculating A_{ij} using mathematical equations for FR and DP. In the AD theory FR and DP are usually transformed to corresponding variables that have functional relationship. For example, an FR_1 is to measure the air velocity, and the DP_1 is a Pitot tube that can measure the stagnation pressure. The FR is represented with the velocity of air and DP is represented by the associated stagnation pressure. Then the value of A_{11} can be calculated with fluid mechanics (Suh 2001). In other cases where the relationship between FR and DP is vague, determining the value of A_{ij} requires systematic consideration rather than subjective judgment.

With A_{ij} representing the effect of DP_j on FR_i, when A_{ij} equals to zero, variation of DP_j does not affect FR_i, rendering FR_j and FR_i independent of each other.

But how do we establish the effect of DP_j on FR_i $(i \neq j)$?

In most cases, it depends usually on the experience and judgment of the designer. As discussed in Sect. 3.4, when the state of FR_1 is changed from point A to C, only DP_1 needs to be changed from state $DP_{1(A)}$ to state $DP_{1(C)}$ without affecting FR_2. Herein, the important concept of "state" needs to be elaborated. What are involved in the "state" is vague at present. It means one will find it confusing when one needs to determine whether the state of DP_1 has changed, especially when working on an innovative design. One would have no clue about what needs be considered. Highly skilled designers may get the solution quickly, while less experienced ones may find it difficult and not knowing where to begin. What is needed is a set of comprehensive and well-defined characteristics to describe the FRs and DPs, a task can be easily addressed with the basic-elements in Extenics.

As discussed, there are 7 characteristics for FR (Eq. 5.4) and 3 characteristics for DP (Eq. 5.16) in general and matrix of A_{ij} can be expressed as

$$A_{ij} = \frac{\partial \mathrm{FR}_i}{\partial \mathrm{DP}_j} = \frac{\partial \vec{u}}{\partial \vec{v}} = \begin{bmatrix} \frac{\partial u_1}{\partial v_1}, & \frac{\partial u_1}{\partial v_2}, & \frac{\partial u_1}{\partial v_3} \\ \vdots & \frac{\partial u_m}{\partial v_n}, & \vdots \\ \vdots & \vdots & \vdots \\ \frac{\partial u_7}{\partial v_1}, & \frac{\partial u_7}{\partial v_2}, & \frac{\partial u_7}{\partial v_3} \end{bmatrix} \tag{5.62}$$

The element "$\frac{\partial u_m}{\partial v_n}$" in Eq. (5.62) represents the relationship between the mth characteristic of FR_i (u_m) and the nth characteristic of DP_j (v_n). When v_n has an impact on u_m, $\frac{\partial u_m}{\partial v_n}$ should be nonzero. The opposite is also true. If there exist any elements that are nonzero, A_{ij} is nonzero. In this way, Eq. (5.62) provides various information needed to determine A_{ij}.

When reviewing the characteristics of DP, one sees that a function is a manifestation of configuration and arrangement. If there is any change in the functional performance or output, it must be caused by variations of characteristics pertaining to configuration and arrangement.

As discussed in Sect. 3.4, "independence" is not an absolute independence. Rather, it is conditional and delimited. It means the characteristics vary within a reasonable range. Moreover, changes should be considered in the process when DPs work to satisfy FRs. Note that the state changes of a DP could be proactive changes or reactive changes in the working process. When a DP varies during application process, for example, the natural wear of tyre, the variation is called proactive change. In comparison, reactive changes are caused by certain requirements such as the functional requirement. In such a case, changes are designed and not natural. Since DPs are designed to satisfy the corresponding FRs while be limited by constraints, the reactive changes of the DPs are generally designed to satisfy the changes of the above two factors.

Consequently, a three-steps method for judging the dependence of DP_j and FR_i is proposed as follows.

Step 1: Establish and Clarify all possible state changes of FR_j and Cs.
Step 2: Determine all possible state changes of corresponding DP_j.
Step 3: Judge the effect of DP_j on FR_i ($i \neq j$) following Eq. (5.62).

As discussed in Sect. 5.2.4, constraints can be added to the corresponding basic-element models or embodied in the requirements of characteristics. Thus, constraints can be also incorporated into the basic-element of FR_j or DP_j, making it feasible to judge the effect of DP_j on FR_i ($i \neq j$) with Eq. (5.62).

Example 5.6 A task is given to three designers to design a pair of eyeglasses. The function of eyeglasses is to correct defective vision and to facilitate improved vision. The function can be decomposed to several sub-functions.

FR_1: Correct light direction
FR_2: Position lenses in front of eyeballs properly and steadily

A design scheme is developed as follows (shown in Fig. 5.3)

DP_1: lenses
DP_2: glasses frame

What's the independence of this scheme?

One designer thinks that when someone needs a pair of glasses and goes to the optician, he or she chooses a frame and the lenses with the corresponding diopter would be processed to fit the chosen frame. Thus, the lenses are dependent on the frame.

Another designer, however, has a different opinion that the FR_2 (Keep the lenses in front of eyeballs properly and steadily) has a close relationship with the lenses, since the lenses must be positioned properly to ensure the clarity of vision. Thus, FR_2 is also dependent upon DP_1.

It seems the two designers are all correct. So, it is confusing as to whether there exists any coupling that needs be addressed.

The third designer applies the basic-element model to deal with the design task using the following design scheme:

FR_1: Correct light direction

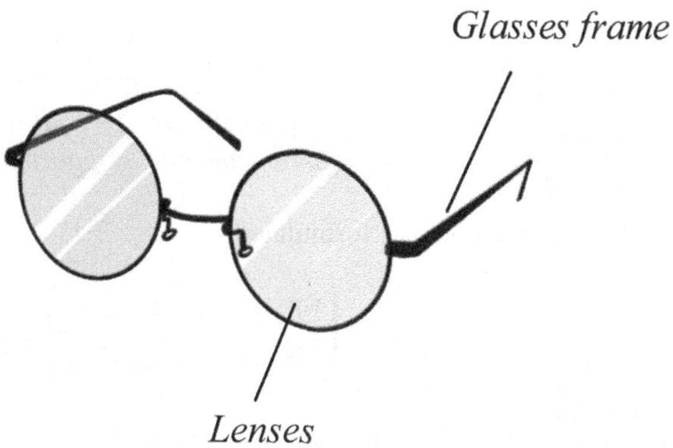

Fig. 5.3 A kind of glasses

$$\mathrm{FR}_1 = \begin{bmatrix} \textit{Correct, dominating object,} & \textit{light direction} \\ \textit{acting object} & \mathrm{DP}_1 \\ \textit{receiving object,} & u_3 \\ \textit{time,} & \textit{when using glasses} \\ \textit{location,} & \textit{through}\,\mathrm{DP}_1 \\ \textit{degree,} & \textit{image clarity} \\ \textit{mode,} & \textit{refraction of light} \end{bmatrix} \tag{5.63}$$

FR_2: Keep lenses in front of eyeballs properly and steadily.

$$\mathrm{FR}_2 = \begin{bmatrix} \textit{Keep, dominating object,} & \textit{lenses} \\ \textit{acting object} & \mathrm{DP}_2 \\ \textit{receiving object,} & \textit{users} \\ \textit{time,} & \textit{when using glasses} \\ \textit{location,} & \textit{in front of eyeballs} \\ \textit{degree,} & \textit{properly and steadily} \\ \textit{mode,} & \textit{supported by ears and nose} \end{bmatrix} \tag{5.64}$$

DP_1: lenses

$$\mathrm{DP}_1 = \begin{bmatrix} \textit{lenses,} & \textit{function,} & \textit{correct light direction} \\ & \textit{configuration,} & \textit{curved piece of lenses} \\ & \textit{arrangement,} & \textit{clamped by glasses frame} \end{bmatrix} \tag{5.65}$$

DP_2: glasses frame

$$\mathrm{DP}_2 = \begin{bmatrix} \textit{glasses frame,} & \textit{function,} & \begin{matrix}\textit{Keep lenses in front of} \\ \textit{eyeballs properly and steadily}\end{matrix} \\ & \textit{configuration,} & \textit{frame structure} \\ & \textit{arrangement,} & \textit{supported by ears and nose} \end{bmatrix} \tag{5.66}$$

The design matrix is formulated as

$$\begin{Bmatrix} \mathrm{FR}_1 \\ \mathrm{FR}_2 \end{Bmatrix} = \begin{bmatrix} X & A_{12} \\ A_{21} & X \end{bmatrix} \begin{Bmatrix} \mathrm{DP}_1 \\ \mathrm{DP}_2 \end{Bmatrix} \tag{5.67}$$

where A_{12} and A_{21} are the parameters to be determined.

(1) Determine the value of A_{12}.

Step 1: Define and establish all the possible state changes of FR_2 and Cs when DP_2 works to satisfy FR_2. When FR_2 functions, the dominating object, receiving object, time, location, and mode are all fixed, while the degree can be changed to "not proper or steady" during the employment.

Step 2: Determine all the possible state changes of the corresponding DP_2.

The change of DP_2 may be deformation that results in FR_2 to be not properly or steadily performed. Herein, the change of DP_2 is a proactive one. The characteristics of configuration and arrangement could induce state changes.

Step 3: Judge the effect of DP_2 on FR_1 according to Eq. (5.62).

The dominating object (light direction), acting object (DP_1), receiving object, time (when using glasses), location (through DP_1), degree (image clarity), and mode (refraction of light) of FR_1 would not be changed when DP_2 has some deformation, i.e.,

$$A_{12} = \frac{\partial \text{FR}_1}{\partial \text{DP}_2} = \frac{\partial \vec{u}}{\partial \vec{v}} = \begin{bmatrix} \frac{\partial u_1}{\partial v_1}, & \frac{\partial u_1}{\partial v_2}, & \frac{\partial u_1}{\partial v_3} \\ \vdots & \frac{\partial u_m}{\partial v_n}, & \vdots \\ \vdots & \vdots & \vdots \\ \frac{\partial u_7}{\partial v_1}, & \frac{\partial u_7}{\partial v_2}, & \frac{\partial u_7}{\partial v_3} \end{bmatrix} = 0 \tag{5.68}$$

(2) Determine the value of A_{21}.

Step 1: Define and establish all possible state changes of FR_1 and Cs when DP_1 works to satisfy FR_1. When FR_1 functions, the dominating object, receiving object, time, location, and mode are all fixed. The degree can be changed to "not clear" during the employment.

Step 2: Determine all possible state changes of the corresponding DP_1.

The image quality depends on the configuration of DP_1. The state changes could be abrasion or scratch, resulting in poor or unclear vision. The change of DP_1 is also a proactive one.

Step 3: Judge the effect of DP_1 on FR_2 according to Eq. (5.62).

The dominating object (the glasses), acting object (DP_2), receiving object (users), time (when using glasses), location (in front of eyeballs), degree (properly and steadily), and mode (supported by ears and nose) of FR_2 would not be changed when DP_1 has some abrasion, i.e.,

$$A_{21} = \frac{\partial \text{FR}_2}{\partial \text{DP}_1} = \frac{\partial \vec{u}}{\partial \vec{v}} = \begin{bmatrix} \frac{\partial u_1}{\partial v_1}, & \frac{\partial u_1}{\partial v_2}, & \frac{\partial u_1}{\partial v_3} \\ \vdots & \frac{\partial u_m}{\partial v_n}, & \vdots \\ \vdots & \vdots & \vdots \\ \frac{\partial u_7}{\partial v_1}, & \frac{\partial u_7}{\partial v_2}, & \frac{\partial u_7}{\partial v_3} \end{bmatrix} = 0 \tag{5.69}$$

The design matrix is

$$\begin{Bmatrix} FR_1 \\ FR_2 \end{Bmatrix} = \begin{bmatrix} X & 0 \\ 0 & X \end{bmatrix} \begin{Bmatrix} DP_1 \\ DP_2 \end{Bmatrix} \tag{5.70}$$

At this point it is evident that the design problem is an uncoupled one.

5.4 Discussion and Summary

The set of innovative design thinking principles discussed in the chapter was a result of synergizing Extenics with the AD theory. The zigzag mapping process and the two AD design axioms provide designers with operation process and criteria. Meanwhile, the basic-element models and extension innovation methods featured in Extenics are viable for solving wicked design problems. It was shown that AD and Extenics are mutually complementary.

The synergy of the two theories provides an approach to reframe a design problem in a structured manner by redefining functional requirements, design parameters, coupling issues and design constraints using basic-element models. Characteristics in basic-elements are determined by performing a comprehensive analysis on the concepts they define. True needs can be easily identified during the process of gathering information and building structured basic-element models. Viable innovative solutions and strategies not immediately apparent can be formulated by following the extension of basic-elements. Furthermore, the relationships between functional and physical domains are specified and the method for judging the dependence of a design is proposed based on the synergy. Consequently, the iterative design process and design criterion are reformulated with the synergized formulation. It was seen that design innovation by such a synergy demonstrated great advantages in fostering design thinking and addressing difficult issues including design coupling and opening up design solution domain.

References

Benavides, E. M. (2011). *Advanced engineering design: An integrated approach.* Woodhead Publishing Limited.

Farid, A. M. & Suh, N. P. (Eds.). (2016). *Axiomatic design in large systems: Complex products, buildings and manufacturing systems.* Springer.

Suh, N. P. (2001). *Axiomatic design: Advances and applications.* New York: Oxford University Press.

Yang, C. Y. & Cai, W. (2013). *Extenics: Theory, method and application.* Beijing: Science Press.

Zhao, Y. W. (2001). Study of conceptual design of extension for mechanical products. *Engineering Science, 3*(5), 67–71.

Chapter 6
Innovative Design Methodology

A natural outcome of the creative synergy of the AD theory with Extenics, the methodology presented in the chapter addresses critical issues such as ambiguity and lack of criteria for guiding design iteration. Innovative thinking is fostered through cognitively reframing a design problem and redefining functional requirements, design parameters, coupling and design constraints using basic-element models. The novel methodology allows a designer to identify the real need, eliminate preconceptions, and develop working solution strategy. By following the principles of design thinking outlined in the previous chapters, the designer is set to explore wide-open design space for innovative design solutions.

6.1 Functional Requirement Analysis

Functional requirement (FR) is the design target that must be satisfied to address the true need. Functional requirements are solution neutral and measurable using design parameters. Determination of functional requirement is critical to the validity of design solution and therefore needs to be taken thoughtfully with care. Not all imposed or given requirements are functional requirements. Design constraints are oftentimes confused with functional requirements, causing confusion in the thinking and decision-making processes. There are three criteria for telling a functional requirement apart from a constraint.

Criterion 1: Functional requirement puts emphasis on the realization of an action whereas constraint is biased towards performance requirement or restriction on solution.

Criterion 2: Functional requirement is required to be satisfied with a corresponding design parameter whereas constraint has certain effects on one or more design parameters.

Criterion 3: In terms of semantic, functional requirement is a verb-noun pair or phrase indicating the action must be taken to meet the requirement. Constraint on

W. Li et al., *Principles of Innovative Design Thinking*,
https://doi.org/10.1007/978-981-19-0485-1_6

the other hand generally carries an adjective or adverb indicating a rigid boundary or envelope that is not to be violated.

Example 6.1 For example, of the 12 requirements listed below, which should be treated as functional requirements, and which should be treated as constraints?

(1) Be low in energy consumption
(2) Be high in brightness
(3) Be small in volume
(4) Be long in service life
(5) Be low in cost
(6) Be minimal in material
(7) Preserve food
(8) Provide space
(9) Support weight
(10) Bear impact
(11) Control speed to less than 20 m/s
(12) Sustain force using the least iron

One can easily determine that the first six requirements are constraints and the 7th to 10th are functional requirements. The 11th and 12th involve both a functional requirement and a constraint. For instance, "Control speed" is the functional requirement while "less than 20 m/s" is the associated constraint. "Sustain force" is the functional requirement and "using the least iron" is the constraint.

After a functional requirement is determined and defined, the decomposition of it is performed with rigorousness. Improperly executed or poorly considered decomposition can have a negative impact on misinterpreting coupling, and not all arbitrarily executed decompositions would lead to or result in an ideal design matrix.

It is intuitive for a designer to identify and decompose functional requirements by considering the time instance at which they are initiated or terminated and the time duration over which they are to stay on or off. For example, "turn on the light" and "turn off the light", or "connect the circuit" and "disconnect the circuit" may inadvertently be treated as pairs of functional requirements by inexperienced designers. While easy to operate, however, this line of decomposition may cause serious problems when analyzing the coupling of different FRs. The reason is that different states of certain components in the design may lead to the exertion of different FRs. When FRs are distinguished in this way, the corresponding DPs would overlap or even be identical. Satisfying different FRs with the same DPs would render an undesirable result—a coupled design matrix. Moreover, couplings caused by overlapping DP are more common than coupled design problems that call for decoupling operation. In designs that involve moving parts, the states of certain components are timed to deliver or stop the function they are designed for. For example, when needed, a key in a keyboard can be pressed down or released, light can be turned on or off, a spur gear can engage or disengage its matching pinion gear. Because it can be in a state of change, defining and decomposing a FR from such a simplistic view is not recommended nor preferred.

As functions are either delivered or performed in orders in which that they are designed to follow, considering the inherent sequential logical order is preferred for identifying and decomposing functional requirements. Logical relationship means that the realization of one function is contingent upon the realization of another function. As an illustrative example, if one wants a light bulb to glow (ignoring the function of turning off the light bulb for now), one must make sure that the function of connecting the circuit is working properly. FRs obtained through decomposition operation that follow logical relationship are satisfied by different components or subsystems. They are inherently different from the FRs obtained through considering the time instance at which they are performed or the time span over which they are activated. Some may argue that logical relationship implies dependency and coupling. For example, one may see the function of illumination of a light bulb as being dependent on the function of connecting the circuit. However, given that they are ubiquitous and unsolvable, logical relationship needs not be addressed in the design process. The confusion could be avoided by properly defining the functional requirements using the following affair-element model

$$\mathrm{FR} = \begin{bmatrix} O_a, \; dominating\,object, & u_1 \\ acting\,object & u_2 \\ receiving\,object, & u_3 \\ time, & u_4 \\ location, & u_5 \\ degree, & u_6 \\ mode, & u_7 \end{bmatrix} \tag{6.1}$$

It is a description of an action from multiple perspectives. The characteristics such as time, location, mode, and degree in the affair-element model are the specific parameters of the FRs. As a FRs is restricted by its parameters, logical relationship can be expressed as a restriction of characteristics in basic-element, and the association with other FRs is embedded in and carried by the restrictions of the FR itself. For example, the glowing function of a light bulb has the restriction of "when electrified" in the characteristic of time, as shown in Eq. (6.2).

$$\mathrm{FR} = \begin{bmatrix} glow, \; dominating\,object, & u_1 \\ acting\,object & the light\,bulb \\ receiving\,object, & u_3 \\ time, & when\,electrified \\ location, & u_5 \\ degree, & u_6 \\ mode, & u_7 \end{bmatrix} \tag{6.2}$$

It implies that the light bulb will glow when electrified, and this quality will not change regardless of the light bulb being electrified or not. Consequently, the glowing function of the light bulb is independent of the function of connecting the circuit.

Even though it is not recommended that FR decomposition be done by considering the time moment at which requirements are met or satisfied, analyzing FRs from the perspective of time is both intuitive and helpful for identifying the logical relationship that underlies all the FRs.

Example 6.2 Consider a general case where one functional requirement is to control the working state of a component devised to facilitate another functional requirement. It requires a time of delivery, t_1, and a time interval, t_2.

Intuitively the FR can be decomposed into.

FR_1: Perform at time $t = t_1$,
FR_2: Stop after time $t = t_1 + t_2$.

However, when determining the corresponding design parameters that underlie the FRs, one finds it difficult to differentiate the associated DPs for the reason that they are timed within the same system. While it is evident that FR_1 and FR_2 are not properly defined, nevertheless, they explicitly indicate the working order of the system. One understands that time is a critical parameter and proper timing is required. Thus, generating time signal is a fundamental functional requirement. Coming from this line of reasoning, an additional FR is identified as controlling the state of electric circuit to facilitate proper timing.

$$FR_1 = \begin{bmatrix} generate, dominating\, object, & signal\, of\, time \\ acting\, object & DP_1 \\ receiving\, object, & u_3 \\ time, & working\, order \\ location, & u_5 \\ degree, & u_6 \\ mode, & u_7 \end{bmatrix} \tag{6.3}$$

$$FR_2 = \begin{bmatrix} control, dominating\, object, & the\, state \\ acting\, object & DP_2 \\ receiving\, object, & electric\, circuit \\ time, & u_4 \\ location, & u_5 \\ degree, & \begin{matrix} respond\, intimeto \\ the\, time\, signal \end{matrix} \\ mode, & u_7 \end{bmatrix} \tag{6.4}$$

A design scheme for a cam-controlled electrical system is shown in Fig. 6.1, where the cam is DP_1 and the lever is DP_2.

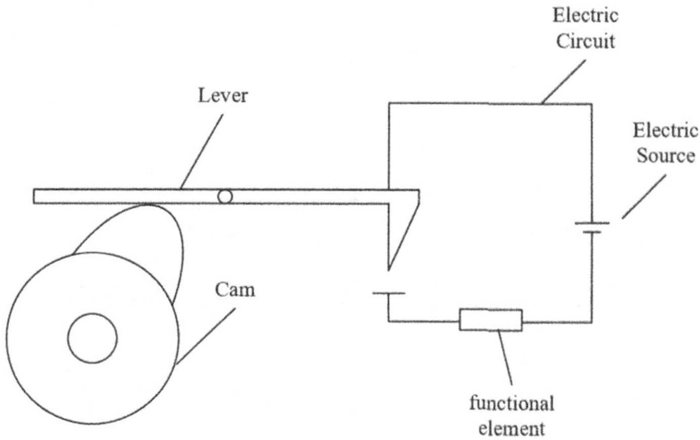

Fig. 6.1 A cam controlled electric system

6.2 Design Parameter Analysis

Determining the DPs that underlie FRs can oftentimes be vague and capricious to the designer. At present this process mainly depends on the designer's subjective judgment and knowledge storage.

According to the concept of design synthesis in AD (Farid and Suh, 2016), the synthesis process reconciles the conceptualization of a set of design parameter primitives **DP** with the satisfaction of a set of FRs as

$$\text{DP} = f_s(\text{FR}, \mathbf{DP}) \tag{6.5}$$

To formalize the thinking process "f_s (·, ·)" and make the design process more manageable, the process of generating DP is defined as the process of solving feasibility problems following Extenics. As discussed in Chap. 4, a feasibility problem in Extenics (Cai, 1994) is to search for the proper value of the given matter-element that belongs to a certain range, i.e., $c_0(O_0) \in V_0$. Then, $\mathbf{DP}_i$ can be expressed with the matter-element model denoted as $\text{M}_i = (\mathbf{DP}_i, c_0, v_0)$. Herein, the "$\mathbf{DP}_i$" is not for a specific design parameter. It is the design parameter primitive that underlie the FR_i. As shown in Eq. (6.5), DP_i are design parameters that are included in $\mathbf{DP}_i$. Consequently, the generation of DP_i is the process of determining the design parameters from the conceptual design parameter primitives $\mathbf{DP}_i$ that satisfy FR_i. Then the implication functions are established to obtain DP_i from the fuzzy descriptions of FR_i. The solution process is divided into several steps as shown in Fig. 6.2.

Step 1. Considering the matter-element model of $\mathbf{DP}_i$, the feasibility problem is defined as

$$M_i = (\mathbf{DP}_i, c_0, v_0), v_0 \in V_0 \tag{6.6}$$

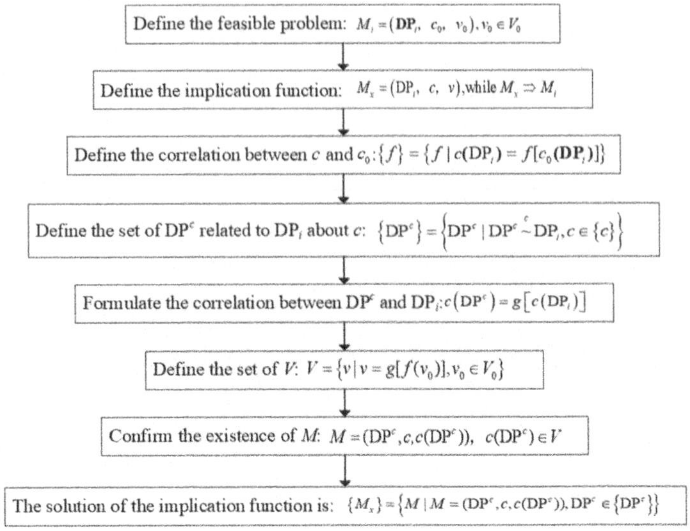

Fig. 6.2 Solution procedure for implication function

where c_0 is the characteristic related to meeting FR_i, v_0 is the value of $\mathbf{DP}_i$ corresponding to the characteristic c_0, and V_0 is the range of value determined by FR_i. To obtain feasible DP_i for the given FR_i is to make the value (v_0) belong to V_0 by solving the implication function.

Step 2. Define the implication function:

$$M_x = (\mathrm{DP}_i, c, v), M_x \Rightarrow M_i \tag{6.7}$$

where DP_i is the proper design parameter being searched for, while c is a set of characteristics related to c_0. To realize M_x, it is necessary to obtain c first. Since c is related to c_0 and c_0 is determined, c can be obtained by performing an analysis of c_0 using Extenics methods such as divergence method, correlative method, implication method, opening-up method, and conjugation method.

Step 3. Set the correlation between c and c_0. The relationship is mathematically expressed as

$$\{f\} = \{f|c(\mathrm{DP}_i) = f[c_0(\mathbf{DP}_i)]\} \tag{6.8}$$

where DP_i is a set of uncertain design parameters with the characteristics of "c", and "f (·)" represents the relationship between $c(\mathrm{DP}_i)$ and $c_0(\mathbf{DP}_\mathrm{i})$, which can be in any form, mathematical or not.

Step 4. Define the set of corresponding objects DP^c related to DP_i about c ($c \in \{c\}$) as follow

$$\{DP^c\} = \left\{DP^c | DP^c \overset{c}{\sim} DP_i, c \in \{c\}\right\} \tag{6.9}$$

where the symbol "$\overset{c}{\sim}$" denotes that DP^c and DP_i are correlative about c.

Step 5. Formulate the correlation between DP^c and DP_i as:

$$c(DP^c) = g[c(DP_i)] \tag{6.10}$$

where "$g(\cdot)$" represents the relationship between $c(DP^c)$ and $c(DP_i)$.

Step 6. Define the set of V, which is the range of value of DP^c

$$V = \{v | v = g[f(v_0)], v_0 \in V_0\} \tag{6.11}$$

where $f(v_0) = f[c_0(DP_i)] = c(DP_i)$, $g[c(DP_i)] = g[f(v_0)]$.

Step 7. Confirm the existence of M with the definition

$$R = (DP^c, c, c(DP^c)), c(DP^c) \in V \tag{6.12}$$

then the expression of DP_i can be obtained as

$$c(DP_i) = g^{-1}(c(DP^c)) \tag{6.13}$$

$$c_0(DP_i) = f^{-1}g^{-1}(c(DP^c)) \in V_0 \tag{6.14}$$

which means that the existence of M can lead to the result that the value (v_0) belongs to V_0, i.e. the realization of M_i.

Step 8. The solution of the implication function M_x is represented as a set in Eq. (6.15).

$$\{M_x\} = \left\{M | M = (DP^c, c, c(DP^c)), DP^c \in \{DP^c\}\right\} \tag{6.15}$$

Discussion

In this book, design parameter (DP) is represented with a matter-element model as given in Eq. (6.16).

$$\mathrm{DP} = \begin{bmatrix} O_m, & function, & v_1 \\ & configuration, & v_2 \\ & arrangement, & v_3 \end{bmatrix} \tag{6.16}$$

where the function of DP is represented as

$$v_1(\mathrm{DP}) = \begin{bmatrix} O_a, & dominating\,object, & u_1 \\ & acting\,object & u_2 \\ & receiving\,object, & u_3 \\ & time, & u_4 \\ & location, & u_5 \\ & degree, & u_6 \\ & mode, & u_7 \end{bmatrix} \tag{6.17}$$

At initial design stage, information is incomplete and generally about the primary functional requirement. Thus, the $\mathbf{DP}_i$ in Step 1 is an abstract concept. In case there is only information about the functional requirement, the characteristic c_0 in Step 1 is the characteristics of function in the matter-element model. The correlative characteristic c in Step 2 could be behavior, action, property, structure, or field, which is relative to concrete parameters v_2 and v_3 (as shown in Eq. (6.16)). It could be some intermediate characteristic, such as property and behavior, as defined in Chap. 5 ($\varphi = g(v_2, v_3)$). And it could also be some extended parameters of v_2 and v_3, such as structure and material. Consequently, the DP_i is more specific with the consideration of characteristic c. In Step 4, the corresponding objects DP^c related to DP_i about c is defined, which is a further step towards the concretization of design parameter.

In the lower layers of decomposition, determination of DP is guided by both the higher layer DP and the corresponding FR, thus the information is more than the first layer of decomposition. The further the decomposition proceeds, the more information would emerge, allowing the progressive converging to the solution space.

Thus it can be summarized that the determination of design parameter is a process of clarifying the basic-element model sequentially by specifying the characteristics of it.

6.3 Mapping Between FRs and DPs

Mapping between FRs and DPs is established by decomposing them gradually. It can be implemented regardless of the design scheme exists or not. When the design task is to obtain a design scheme that is yet to exist, mapping is conducted to specify FRs and DPs by decomposing them. If there exists a design scheme that is related to the FRs, the mapping process facilitates the evaluation and evolution of the scheme.

Mapping results in a design matrix that makes explicit the inherent coupling problem. Design coupling can be addressed and resolved using Extenics based techniques to be introduced in detail in the next chapter. Decomposition helps to define FRs and identify critical characteristics, allowing a problem to be reframed and transformed to enable innovation.

Mapping-enabled reframing is a solution-based design thinking approach that investigates functional requirement along with considering the corresponding DP. As discussed in Chap. 5, FR_i is represented using an affair-element model with the corresponding DP_i included as follows

$$\mathrm{FR}_i = \begin{bmatrix} O_a, & dominating\,object, & u_1 \\ & acting\,object & \mathrm{DP}_i \\ & receiving\,object, & u_3 \\ & time, & u_4 \\ & location, & u_5 \\ & degree, & u_6 \\ & mode, & u_7 \end{bmatrix} \tag{6.18}$$

The matter-element model of DP_i with n functional characteristics is denoted as

$$\mathrm{DP}_i = \begin{bmatrix} O_m, & function_1, & v_{11} \\ & function_2, & v_{12} \\ & \vdots & \vdots \\ & function_n, & v_{1n} \\ & configuration, & v_2 \\ & arrangement, & v_3 \end{bmatrix} \tag{6.19}$$

where "v_{1n} (DP_i)" represents the n-th functional characteristic of DP_i, which is expressed as an affair-element model:

$$v_{1n}\,(\mathrm{DP}_i) = \begin{bmatrix} O_a, & dominating\,object, & u_1 \\ & acting\,object, & u_2 \\ & receiving\,object, & u_3 \\ & time, & u_4 \\ & location, & u_5 \\ & degree, & u_6 \\ & mode, & u_7 \end{bmatrix} \tag{6.20}$$

The functional characteristics of DP_i are the actual functions satisfying the functional requirements, and the relation between them is denoted as

$$\mathrm{FR}_{in} = T v_{1n}(\mathrm{DP}_i) \tag{6.21}$$

The relationship between functional requirement and design parameter is detailed and refined by analyzing the relationships between the characteristics. The synergized formulation solves the problem of lacking unified and detailed expression of elements in the AD theory, with the basic-element of Extenics providing the specific characteristics in ordered triples for description. In addition, expressions with basic element are well classified from the semantic viewpoint, making explicit the relationship between FRs and DPs. The formulation with basic-element model involves abundant information that can be effectively deployed to promote the mapping process.

According to AD, a design is considered an ideal design if each off-diagonal element is zero, i.e., $A_{ij} = 0\ (i \neq j)$. It means that FR_i is completely defined by DP_i. In other words, DP_i is supposed to be the subject of the verb phrase of FR_i. In that case that the design cannot be an ideal design, the off-diagonal elements should be much smaller than the diagonal elements to achieve a robust design, i.e., $A_{ii} > > A_{ij}$, meaning the corresponding DP_i is the main design parameter that satisfies FR_i (Suh, 2001). During the design process, trial-and-error approach will risk making the off-diagonal elements too great. With the synergized formulation, DP_i is set to be the acting object characteristic of FR_i, which represents the subject of the verb phrase of FR_i. In such manner, the DP_i corresponding to FR_i can be properly determined.

For example, consider the design of the hot and cold water faucet, which is a typical illustration in AD, the functional requirements are:

FR_1: Control water flow rate Q.
FR_2: Control water temperature T.

It is intuitive for a designer, both inexperienced and skilled alike, to come up with the following design scheme (Farid and Suh, 2016; Suh, 2001) as shown in Fig. 6.3.

DP_1: A knob to control flow rate of hot water.
DP_2: A knob to control flow rate of cold water.

The design function is therefore

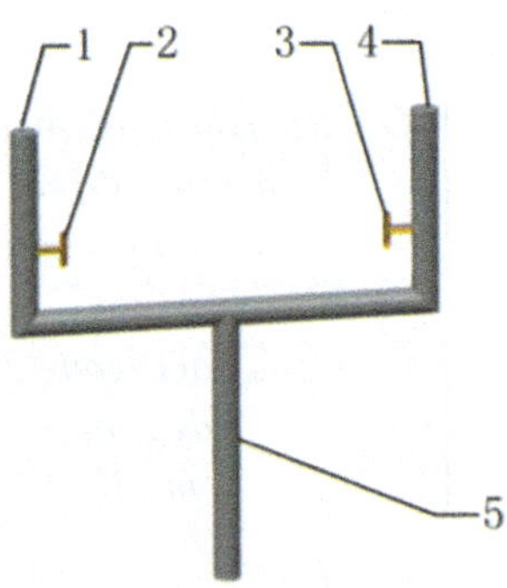

1- Cold water line 2- Knob A 3- Knob B 4- Hot water line 5- Outlet pipe

Fig. 6.3 The coupled hot water and cold water faucet

$$\begin{pmatrix} FR_1 \\ FR_2 \end{pmatrix} = \begin{bmatrix} A_{11} & A_{12} \\ A_{21} & A_{22} \end{bmatrix} \begin{pmatrix} DP_1 \\ DP_2 \end{pmatrix} \tag{6.22}$$

which is evidently a coupled design.

As a comparison, the design process is reframed and the FR_i can be expressed as

$$FR_1 = \begin{bmatrix} control,\ dominating\ object, & Q \\ acting\ object & DP_1 \\ receiving\ object, & water \\ time, & u_4 \\ location, & u_5 \\ degree, & accurately \\ mode, & turn\ on\ (off) \\ tool, & u_8 \end{bmatrix} \tag{6.23}$$

$$FR_2 = \begin{bmatrix} control,\ dominating\ object, & T \\ acting\ object & DP_2 \\ receiving\ object, & water \\ time, & u_4 \\ location, & u_5 \\ degree, & accurately \\ mode, & turn\ on\ (off) \\ tool, & u_8 \end{bmatrix} \tag{6.24}$$

The information provided by FRs clarifies the direction of searching for the proper solution. Intuition-based impulse and preconception for the design scheme are replaced by solution-based innovative ideas. The DPs are defined as

$$DP_1 = \begin{bmatrix} faucet, & function, & \begin{matrix} control\ of\ the\ water \\ flow\ rate\ accurately \end{matrix} \\ & configuration, & v_2 \\ & arrangement, & v_3 \end{bmatrix} \tag{6.25}$$

$$DP_2 = \begin{bmatrix} faucet, & function, & \begin{matrix} control\ of\ the\ water \\ temperature\ accurately \end{matrix} \\ & configuration, & v_2 \\ & arrangement, & v_3 \end{bmatrix} \tag{6.26}$$

This means that DP_1 is a faucet for the accurate control of the water flow rate, while DP_2 is a faucet for the accurate control of the temperature.

For example, Fig. 6.4 displays a configuration of the design obtained with ease using the formulations above.

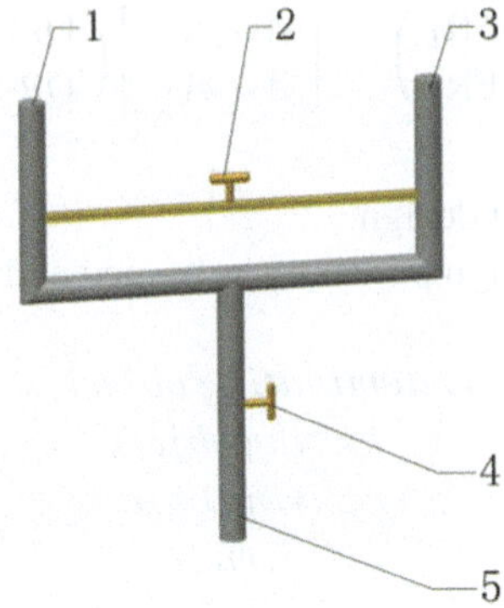

1- Cold water line 2- Knob A 3- Hot water line 4- Knob B 5- Outlet pipe

Fig. 6.4 The uncoupled hot water and cold water faucet

As expected, DP_i is the subject of FR_i and FR_i is completely defined by DP_i. The design function is therefore

$$\begin{pmatrix} FR_1 \\ FR_2 \end{pmatrix} = \begin{bmatrix} A_{11} & 0 \\ 0 & A_{22} \end{bmatrix} \begin{pmatrix} DP_1 \\ DP_2 \end{pmatrix} \tag{6.27}$$

Based on the redefined FRs and DPs, the zigzag mapping process is reframed. The primary mapping process, especially the mapping from FR_0 to DP_0, is the most important step to generate the whole design scheme and at the same time the step is also the most obscured due to lack of information. Applying feasibility problem solving evidently contributes in a significant way to the mapping process by transforming vagueness and imprecision into an unequivocal correspondence described by definitive elements. Once DP_0 is determined, one then goes through a process where one zigzags between domains to complete the whole design. As the decomposition process iterates further, more relevant information are available, and the true functional requirement along with the associated design parameter emerges. Consequently, DPs and FRs are obtained by extracting the viable information of the known conditions.

In response to the issue a comprehensive Extenics based method of decomposition is presented in the followings. With affair-element representing FRs and matter-element representing DPs, the decomposition process is ordered as follows:

(1) Identify the true need by abstraction and build the affair-element model of FR_0.
(2) If the description of FR_0 is fuzzy and information is scarce, the feasibility problem model is established.
(3) Define and solve the implication functions to obtain the corresponding DP_0.
(4) Extract x independent functions of DP_i and build corresponding matter-element model.

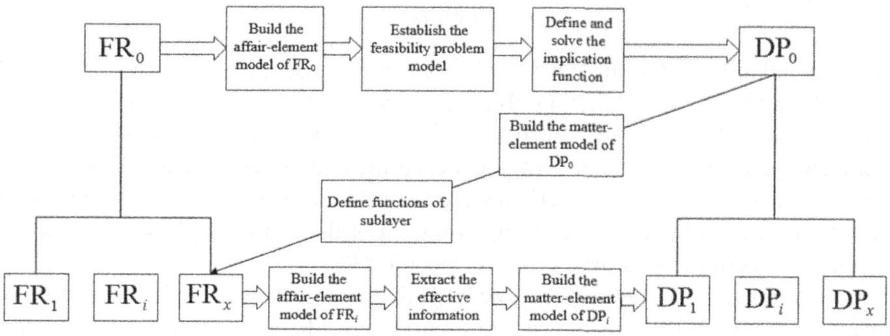

Fig. 6.5 Revised iterative decomposition process

(5) Build corresponding affair-element model of FR_{im} ($m \in (1, x)$) according to each independent function of DP_i.
(6) Extract effective information from the affair-element model of FR_{im}.
(7) Determine the corresponding DP_{im} ($m \in (1, x)$). The feasibility problem solving method can be applied when applicable.
(8) Document design functions generated in the mapping process and make sure they are independent of each other.
(9) Repeat the above steps till the decomposition of each layer is complete.

The revised iterative decomposition process is summarized in Fig. 6.5.

6.4 Expanding Solutions

During the decomposition process in AD, designers tend to select a specific solution for each design parameter. This approach can benefit the determination of the design scheme. However, it also restricts the generation of novel design schemes. In this book, the FRs in the form of affair-element and the DPs in the form of matter-element are abstract concepts instead of concrete structures. In addition affair-element and matter-element are defined with the definite basic characteristics, which can be used as the objectives for extension. According to the principle of divergence, one's thinking mode can be extended. With the tool of transformations, a given basic-element can be extended to multiple basic-elements.

When the existing basic-element cannot solve the problem, extended basic-elements with different objects (O), characteristics(c) or values (v) can be considered. By so doing, the solutions of functional requirements (FRs) and design parameters (DPs) can be expanded.

(1) Extension of the affair-element model of FR

As discussed in Sect. 3.1, the functional requirement is described in terms of ordered triples. Dominating object, acting object and receiving object are the basic components of the verb phrase to describe the functional requirement among the basic characteristics, while the others provide the detailed descriptions with all the aspects of it. In practice, some characteristics are determined while others may be vague for the moment. Divergence and transformation of the characteristics in the FR are helpful in searching for similar or more specific FRs.

For example, a different FR_i' can be extended from FR_i with a different value of the mode. It means FR_i' has the same function with FR_i and the time, location and degree of the function are also the same. Nevertheless, FR_i' is realized in a different mode.

Example 6.3 In the initial stage of design, a functional requirement is to control the light, which is represented as:

$$FR_0 = \begin{bmatrix} control, & dominating\ object, & the\ light \\ & acting\ object & DP_0 \\ & receiving\ object, & u_3 \\ & time, & when\ needed \\ & location, & \begin{matrix} on\ the\ body \\ of\ the\ light \end{matrix} \\ & degree, & accurately \\ & mode, & u_7 \end{bmatrix} \tag{6.28}$$

It means DP_0 is designed to control the light when needed, and the location of the function is in the body of the light. In this case, controlling the light is the main functional requirement while the location of it is determined based on experience. The characteristic of location can be diverged to eliminate preconception, as shown in Table 6.1.

When the substitution transformation is implemented, the FR_0 can be transformed to FR_0^1, FR_0^2 or FR_0^3:

Table 6.1 Divergence of the location characteristic

Object	Characteristic	Value	Extended values
Control	Location	In the body	Around the light, long-range

$$\mathrm{FR}_0^1 = \begin{bmatrix} control,\ dominating\ object, & the\ light \\ acting\ object & \mathrm{DP}_0 \\ receiving\ object, & u_3 \\ time, & when\ needed \\ location, & aroundthelight \\ degree, & accurately \\ mode, & u_7 \end{bmatrix} \tag{6.29}$$

$$\mathrm{FR}_0^2 = \begin{bmatrix} control,\ dominating\ object, & the\ light \\ acting\ object & \mathrm{DP}_0 \\ receiving\ object, & u_3 \\ time, & when\ needed \\ location, & long-range \\ degree, & accurately \\ mode, & u_7 \end{bmatrix} \tag{6.30}$$

$$\mathrm{FR}_0^3 = \begin{bmatrix} control,\ dominating\ object, & the\ light \\ acting\ object & \mathrm{DP}_0 \\ receiving\ object, & u_3 \\ time, & when\ needed \\ location, & \begin{matrix} around\ the\ light, \\ and\ long-range \end{matrix} \\ degree, & accurately \\ mode, & u_7 \end{bmatrix} \tag{6.31}$$

When the increasing transformation is implemented, the FR_0 can be transformed to FR_0^4, FR_0^5 or FR_0^6:

$$\mathrm{FR}_0^4 = \begin{bmatrix} control,\ dominating\ object, & the\ light \\ acting\ object & \mathrm{DP}_0 \\ receiving\ object, & u_3 \\ time, & when\ needed \\ location, & \begin{matrix} on\ the\ body,\ and \\ aroundthelight \end{matrix} \\ degree, & accurately \\ mode, & u_7 \end{bmatrix} \tag{6.32}$$

$$\mathrm{FR}_0^5 = \begin{bmatrix} control, \ dominating\,object, & the\,light \\ acting\,object & \mathrm{DP}_0 \\ receiving\,object, & u_3 \\ time, & when\,needed \\ location, & \begin{matrix} on\,the\,body,\,and \\ long-range \end{matrix} \\ degree, & accurately \\ mode, & u_7 \end{bmatrix} \tag{6.33}$$

$$\mathrm{FR}_0^6 = \begin{bmatrix} control, \ dominating\,object, & the\,light \\ acting\,object & \mathrm{DP}_0 \\ receiving\,object, & u_3 \\ time, & when\,needed \\ location, & \begin{matrix} on\,the\,body, \\ around\,the\,light, \\ and\,long-range \end{matrix} \\ degree, & accurately \\ mode, & u_7 \end{bmatrix} \tag{6.34}$$

Among the six extended functional requirements, one can determine the real requirement concerning about the actual application.

(2) Extension of the matter-element model of DP

The DP is represented with a matter-element as

$$\mathrm{DP} = \begin{bmatrix} O_m, & function, & v_1 \\ & configuration, & v_2 \\ & arrangement, & v_3 \end{bmatrix} \tag{6.35}$$

Each characteristic can be extended to a series of relative attributes. As discussed in Sect. 5.2.2, the characteristics of configuration and arrangement can be diverged as

$$(O_m, configuration, v) -| \begin{Bmatrix} (O_m, shape, v), \\ (O_m, structure, v), \\ (O_m, size, v), \\ (O_m, material, v), \\ \vdots \end{Bmatrix} \tag{6.36}$$

$$(O_m, arrangement, v) -| \begin{Bmatrix} (O_m, position, v), \\ (O_m, assembly\,mode, v), \\ \vdots \end{Bmatrix} \tag{6.37}$$

As long as the function of DP_i remains decomposable into several sub-functions, the decomposition process is incomplete. As a result, the DP_i is an aggregate of sub design parameters. And generally, the details of DP_i cannot be determined yet, and the function, configuration, and arrangement of DP_i are vague.

When the function of DP_i cannot be decomposed further, one can determine the concrete parameters of DP_i. The process of determining the sub-parameters in a DP is a process going from abstraction to specification. During this process, the sub-parameters can be diverged and extended. In so doing, the design solution space is significantly broadened.

Example 6.4 In Example 5.3, the three DPs to satisfy FR_1 (Support a person's weight on buttock), FR_2 (Support a person's back), and FR_3 (Keep a person above the ground) are expressed with matter-elements as:

$$DP_1 = \begin{bmatrix} The\,seat, & function, & \begin{matrix} Support\,a\,person's \\ weight\,on\,buttock \end{matrix} \\ & configuration, & Surface \\ & arrangement, & Horizontal \end{bmatrix} \tag{6.38}$$

$$DP_2 = \begin{bmatrix} The\,back, & function, & Support\,a\,person's\,back \\ & configuration, & Surface \\ & arrangement, & Vertical \end{bmatrix} \tag{6.39}$$

$$DP_3 = \begin{bmatrix} The\,leg, & function, & Keep\,a\,person\,above\,the\,ground \\ & configuration, & Column \\ & arrangement, & Vertical \end{bmatrix} \tag{6.40}$$

And the configuration of DP_1 is extended to shape, size, and material as the sub-parameters of the seat,

$$(the\,seat, configuration, surface) -| \begin{Bmatrix} (the\,seat, shape, v), \\ (the\,seat, size, v), \\ (the\,seat, material, v) \end{Bmatrix} \tag{6.41}$$

Each of the sub-parameters has multiple values, some of which are listed in Table 6.2 for illustration.

Table 6.2 Sub-parameters of the configuration of the seat

Object	Characteristic	Value
Seat	shape	Square, trapezoid, circle
	size	0.2m^2, $0.17\ \text{m}^2$
	material	Fabric, wood, iron

Table 6.3 Sub-parameters of seat arrangement

Object	Characteristic	Value
Seat	Arrangement	Horizontal arrangement, with an angle of 5°

Moreover, the arrangement of DP_1 can also be extended as:

$$\begin{pmatrix} the\,seat, arrangement, horizontally \\ arranged \end{pmatrix} -| \begin{pmatrix} the\,seat, arrangement, with\,an\,angle \\ of\,5\,degree \end{pmatrix} \tag{6.42}$$

Herein, the two arrangements of DP_1 are listed in Table 6.3,

It is evident that the listed values of the four characteristics result in 36 ($3 \times 2 \times 3 \times 2$) possible combinations, i.e., 36 possible design solutions for DP_1, such as a fabric seat in the shape of a square with an area of $0.2m^2$ that is horizontally arranged, a fabric seat in the shape of a trapezoid with an area of $0.2m^2$ that is horizontally arranged, a fabric seat in the shape of a circle with an area of $0.2m^2$ that is horizontally arranged, and so on. Some design solutions are illustrated in Table 6.4, demonstrating different materials and configurations.

The characteristics in DP_2 and DP_3 can be extended similarly. Evidently, the design solution space is very broad and numerous solutions can be obtained with the extensions of the parameters in the matter-element models of DPs.

The decomposition process will eventually lead to a set of design schemes. However, not all of them are feasible. An optimal scheme will emerge once the imposed constraint conditions and requirements are considered and the Independence Axiom is applied.

6.5 Illustrative Example

Corn harvester header is one of the core components that dictates the performance of a harvester with the rate of grain loss and breakage. The energy consumption of the header during the picking process is high. The header of the four-row, self-propelled corn harvester consumes about 17% of the operation energy. Innovative header structure design is commonly explored to seek a better performance in energy efficiency for the corn harvester design. The current design approach renders ambiguity and confusion when searching for design solutions and risking discouraging broad-base thinking that fosters innovation. The knowledge and experience of the designer dominated the process.

The corn plant is shown in Fig. 6.6, and function of a corn harvester is to pick corn ears from cornstalks and then gather corn ears together (Sun et al., 2018) through separating corn ears from the cornstalk at the corn ear stalk.

Table 6.4 Some design solutions for the seat

 A fabric seat in square shape	 A fabric seat in trapezoidal shape	 A fabric seat in circular shape
 An iron seat in square shape	 An iron seat in trapezoidal shape	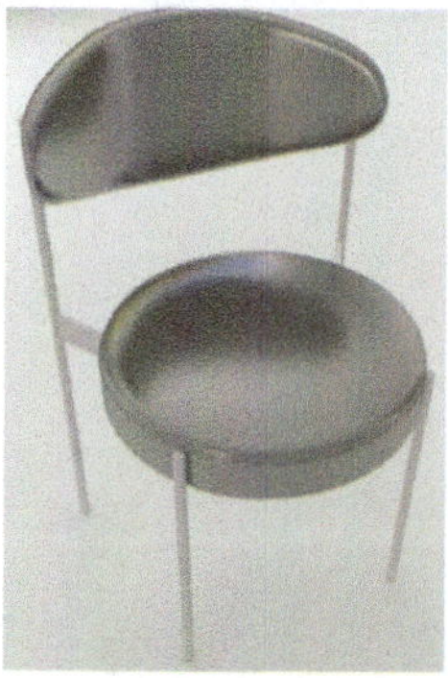 An iron seat in circular shape
 A wood seat in square shape	 A wood seat in trapezoidal shape	 A wood seat in circular shape

Fig. 6.6 Corn plant

Fig. 6.7 A kind of corn harvester header

At present, the main corn harvester headers are all with the key component of snapping roller which is used to convey the corn stalks by rolling. A commonly used harvester header is shown in Fig. 6.7 which is a John Deere product. The corn stalks move towards backward and downward relative to snapping rollers when they enter the gap of two snapping rollers. Since corn ears are too thick to get in the gap, they are pulled off from the cornstalks.

An innovative corn harvester header design is discussed in this section by applying the improved design methodology elaborated in the previous sections. This is presented as a demonstration of the general applicability of the methodology. Having a definitive rule for guiding the mapping process is particularly critical when one is engaged in the design decomposition layers.

6.5.1 Conceptual Design of Corn Harvester Header

In the followings a novel corn harvester header is developed. The improved design procedure is followed to generate the design steps outlined below from which an optimal design configuration is emerged.

Step 1: Define the feasibility problem: $M_0 = (\mathrm{DP}_0, c_0, v_0), v_0 \in V_0$.

It is noted that the "$\mathbf{DP}_0$" herein is not for a specific corn harvester configuration. It is a general designation for the feasible design parameters of FR_0.

The functional requirement (FR_0) of the corn harvester header is harvesting corn ears. The example is presented in such a way that the requirements are set qualitatively rather than quantitatively as shown in Eq. (6.43), where the FR_0 is required to be "with less energy and damage rate".

$$\mathrm{FR}_0 = \begin{bmatrix} harvest,\ dominating\ object, & cornear \\ acting\ object & \mathbf{DP}_0 \\ receiving\ object, & whole-plant\ corn \\ time, & u_4 \\ location, & u_5 \\ degree, & with\ less\ energy\ and\ damage\ rate \\ mode, & pick\ and\ gather\ the\ corns \end{bmatrix} \tag{6.43}$$

"c_0" is divided into two parts: picking corn ears (c_{01}) and gathering the picked corn ears (c_{02}). "v_0" represents the degree of c_0, which is, "with less energy and damage rate". The feasibility problem is represented as Eqs. (6.44) and (6.45).

$$M_0 = \begin{bmatrix} \mathrm{DP}_0, & c_{01}, & v_{01} \\ & c_{02}, & v_{02} \end{bmatrix}, v_0 \in V_0 \tag{6.44}$$

$$c_0 = \begin{pmatrix} c_{01} \\ c_{02} \end{pmatrix} = \begin{pmatrix} picking\ corn\ ears \\ gathering\ corn\ ears \end{pmatrix} \tag{6.45}$$

c_{01} ($\mathbf{DP}_0$) = The function "picking corn ears" of the header.

c_{02} ($\mathbf{DP}_0$) = The function "gathering corn ears" of the header

$$v_0 = \begin{pmatrix} v_{01} \\ v_{02} \end{pmatrix}, V_0 = \begin{pmatrix} with\ less\ energy\ and\ damage\ rate \\ with\ less\ energy\ and\ damage\ rate \end{pmatrix} \tag{6.46}$$

Note that the V_0 is the value range determined by FR_0, since the degree of FR_0 is set qualitatively as "with less energy and damage rate", the V_0 is also set as a qualitative range. To obtain the feasible DP_0 for the given FR_0 is to make the value (v_0) belong to V_0, that is, the functions "picking corn ears" as well as "gathering corn ears" of the header should with less energy and damage rate.

Step 2: Define the implication function: $M_x = (\mathrm{DP}_0, c, v), M_x \Rightarrow M_0, v \in V$ by determining the set of characteristics $\{c\}$ that related to c_0. We know that c is related to c_0 and c_0 is determined by c. Herein, the functions in $\{c_0\}$ are implemented with forces. The corresponding forces are the characteristics related to c_0.

$$c = \begin{pmatrix} c_1 \\ c_2 \end{pmatrix} = \begin{pmatrix} corn\ ears\ picking force \\ corn\ ears\ gathering force \end{pmatrix} \tag{6.47}$$

c_1 (DP_0) = Corn ears picking force of the header.

c_2 (DP_0) = Corn ears gathering force of the header.

According to V_0, corn picking force and corn gathering force should make sure that the corns are harvested with less energy and damage rate. The forces should not be too small to achieve the functions or too great to cause energy waste and high damage rate.

Step 3: Set the correlation between c and c_0: $\{f\} = \{f|c(DP_i) = f[c_0(\mathbf{DP}_i)]\}$. Since there are two pairs of c and c_0, there are also two elements in $\{f\}$.

$$\begin{aligned} \{f\} &= \{f_1, f_2\} \\ c_1(DP_0) &= f_1[c_{01}(DP_0)] \\ c_2(DP_0) &= f_2[c_{02}(DP_0)] \end{aligned} \tag{6.48}$$

The equation "$c_i(DP_0)=f_i[c_{0i}(DP_0)]$"means "$c_i(DP_0)$ is the approach to satisfy $c_{0i}(DP_0)$".

Step 4: Define the set of corresponding objects DP^c related to DP_0 about c ($c \in \{c\}$) as $\{DP^c\} = \left\{DP^c | DP^c \overset{c}{\sim} DP_0, c \in \{c\}\right\}$.

Since c is defined in Eq. (6.47), the DP^c is a set of objects that can provide corn ears picking force and gathering force, denoted as:

$$\left(DP_1^c,\ cornear\ picking\ force,\ v_1\right),\quad \left(DP_2^c,\ cornear\ gathering\ force,\ v_2\right)$$

Herein, the v_1 and v_2 are the concrete forms of forces.

Considering the possibility of expanding solutions from the basic-element expressions as discussed in Sect. 6.4, we carry on the divergence of the values according to one characteristic to help the designer broaden the design space.

$$\begin{aligned} &\left(DP_1^c,\ cornear\ picking\ force,\ v_1\right) \\ &-| \begin{pmatrix} DP_1^c, & cornear\ picking\ force, & pull \\ & & cut \\ & & shear \\ & & break \end{pmatrix} \end{aligned} \tag{6.49}$$

$$\begin{aligned} &\left(DP_2^c,\ cornear\ gathering\ force,\ v_2\right) \\ &-| \begin{pmatrix} DP_2^c, & cornear\ gathering\ force, & push \\ & & lift \\ & & fall \end{pmatrix} \end{aligned} \tag{6.50}$$

Table 6.5 Solution space of conceptual design

Corn picking Corn gathering	Pull	Cut	Shear	Break
Push	Pull off corn ears then push them together	Cut off corn ears then push them together	Shear off corn ears then push them together	Break off corn ears then push them together
Lift	Pull off corn ears then lift them together	Cut off corn ears then lift them together	Shear off corn ears then lift them together	Break off corn ears then lift them together
Fall	Pull off corn ears then make them fall together	Cut off corn ears then make them fall together	Shear off corn ears then make them fall together	Break off corn ears then make them fall together

Herein, four patterns of corn ears picking force and three patterns of corn ears gathering force are established according to expertise. There will be "4×3" kinds of solutions accordingly (Table 6.5).

Step 5: Formulate the correlation between DP^c and DP_0 as:

$$c\left(\mathrm{DP}^c\right) = g[c(\mathrm{DP}_0)] \tag{6.51}$$

where "$g(\cdot)$" represents the relationship between $c(\mathrm{DP}^c)$ and $c(\mathrm{DP}_0)$, that is

$$\begin{aligned} &\{g\} = \{g_1, g_2\} \\ &c_1\left(\mathrm{DP}_1^c\right) = g_1[c_1(\mathrm{DP}_0)] \\ &c_2\left(\mathrm{DP}_2^c\right) = g_2[c_2(\mathrm{DP}_0)] \end{aligned} \tag{6.52}$$

The equation "$c_i(\mathrm{DP}_i^c) = g_i[c_i(\mathrm{DP}_0)]$" means "$\mathrm{DP}_i^c$ is the specific object of DP_0 about c_i.

Step 6: Define the set of V, which is the value range of DP^c.

$$V = \{v | v = g[f(v_0)], v_0 \in V_0\} \tag{6.53}$$

As discussed in Eq. (6.46), V_0 indicates that the functions "pick corn ears" as well as "gather corn ears" of the header should be with less energy and damage rate. Thus, the picking force should be larger than the fracture stress of the ear stalk but not too large considering the energy consumed. The gathering force is to collect corn ears without damaging them, and the energy-consuming should also be considered. As a result, the forces should not be too small to achieve the functions or too great to cause energy waste and high damage rate. Consequently, V can be expressed qualitatively as $V = \begin{pmatrix} proper \\ proper \end{pmatrix}$ for simple.

Accordingly, the solution is selected with the characteristic-element:

$$\left(\mathrm{DP}_1^c, \textit{cornear picking force}, \textit{break}\right), \quad \left(\mathrm{DP}_2^c, \textit{cornear gathering force}, \textit{fall}\right)$$

Step 7: Confirm the existence of M with the definition

$$M = (\mathrm{DP}^c, c, c(\mathrm{DP}^c)), c(\mathrm{DP}^c) \in V \tag{6.54}$$

Obviously, the $\mathrm{DP}^c{}_1$ and $\mathrm{DP}^c{}_2$ can be realized technically. And the existence of M can lead to the result that the value (v_0) belongs to V_0, i.e. the realization of M_i.

Step 8: The solution of the implication function M_x is represented as $\{M_x\} = \{M | M = (\mathrm{DP}^c, c, c(\mathrm{DP}^c)), \mathrm{DP}^c \in \{\mathrm{DP}^c\}\}$.

The DP^c is processed by merging $\mathrm{DP}^c{}_1$ and $\mathrm{DP}^c{}_2$, as:

$$M = \begin{bmatrix} \mathrm{DP}^c, & \textit{corn ear picking}, & \textit{break} \\ & \textit{corn ear gathering}, & \textit{fall} \end{bmatrix} \tag{6.55}$$

Herein, DP^c is a new type of corn harvester header which can break corns and gather them when they fall. According to the implication principle of Extenics, the existence of M will lead to realization of M_0. Consequently, the DP_0 in the physical domain can be found as

$$\mathrm{DP}_0 = \begin{bmatrix} \begin{matrix} \textit{new}-\textit{type corn} \\ \textit{harvester header}, \end{matrix} & \textit{function1}, & \begin{matrix} \textit{corn ears} \\ \textit{picking} \end{matrix} \\ & \textit{function2}, & \begin{matrix} \textit{corn ears} \\ \textit{gathering} \end{matrix} \\ & \textit{configuration}, & v_3 \\ & \textit{arrangement}, & v_4 \end{bmatrix} \tag{6.56}$$

$$\textit{function1} = \begin{bmatrix} \textit{pick}, \textit{dominating object}, & \textit{corn ears} \\ \textit{acting object} & \mathrm{DP}_0 \\ \textit{receiving object}, & u_3 \\ \textit{time}, & \begin{matrix} \textit{when ear stalk arrives} \\ \textit{at the work spot} \end{matrix} \\ \textit{location}, & \textit{ear stalk} \\ \textit{degree}, & \textit{completely detached} \\ \textit{mode}, & \textit{break} \end{bmatrix} \tag{6.57}$$

$$function2 = \begin{bmatrix} gather, dominating\,object, & cornears \\ acting\,object & \mathrm{DP}_0 \\ receiving\,object, & u_3 \\ time, & \begin{matrix} after\,the\,corn\,ears \\ have\,been\,picked\,off \end{matrix} \\ location, & under\,the\,corn\,ears \\ degree, & completely\,gathered \\ mode, & \begin{matrix} catch\,the\,falling\,corn \\ ears\,and\,gather\,them \end{matrix} \end{bmatrix} \quad (6.58)$$

In this section, 12 kinds of solutions are generated among which the most proper one is selected. It would be very difficult to follow AD to generate such a large amount of solutions and obtain the most proper one in such a short amount of time.

6.5.2 Zigzag-Mapping Process

A detail corn harvester header design can be obtained by establishing the zigzagging mapping between the function domain and physical domain using the steps outlined below:

Step 1: Extract x independent functions of DP_0 and build the corresponding matter-element model.

DP_0 is still in form of concept rather than concrete structure. Thus, the characteristics besides the function in matter-element are uncertain, which is shown in Eq. (6.56).

Step 2: Build the corresponding affair-element model of FR_m $(m \in (1, x))$ according to each independent function of DP_0.

According to the expressions in Eqs. (6.57) and (6.58), FR_1 is picking corn ears while FR_2 is gathering corn ears. The affair-element models of FR_1 and FR_2 are:

$$\mathrm{FR}_1 = function1 = \begin{bmatrix} pick, dominating\,object, & corn\,ears \\ acting\,object & \mathrm{DP}_1 \\ receiving\,object, & u_3 \\ time, & \begin{matrix} when\,ear\,stalk\,arrives \\ at\,the\,work\,spot \end{matrix} \\ location, & ear\,stalk \\ degree, & completely\,detached \\ mode, & break \end{bmatrix} \quad (6.59)$$

$$\mathrm{FR}_2 = function2 = \begin{bmatrix} gather, & dominating\ object, & corn\ ears \\ & acting\ object & \mathrm{DP}_2 \\ & receiving\ object, & u_3 \\ & time, & \begin{array}{c} after\ the\ corn\ ears \\ have\ been\ picked\ off \end{array} \\ & location, & under\ the\ corn\ ears \\ & degree, & completely\ gathered \\ & mode, & \begin{array}{c} catch\ the\ falling\ corn \\ ears\ and\ gather\ them \end{array} \end{bmatrix} \tag{6.60}$$

Step 3: Extract the effective information of the affair-element model of FR_{im}, and determine the corresponding DP_{im} ($m \in (1, x)$).

When corn ears reach the work spot of the header, DP_1 provides a top-down force on them. It means DP_1 is a separating unit with the functions of transporting cornstalks and detaching the corn ears. Meanwhile, the separated cornstalks must be supported to make the moving smooth. After corn ears are picked off, they fall off and be caught and gathered by DP_2. The corresponding DP_2 therefore is a device for catching and gathering.

$$\mathrm{DP}_1 = \begin{pmatrix} separating\ unit, & function1, & transfer\ the\ cornstalk \\ & function2, & detach\ the\ corn\ ears \\ & function3, & \begin{array}{c} support\ the\ moving \\ cornstalk \end{array} \\ & configuration, & v_4 \\ & arrangement, & v_5 \end{pmatrix} \tag{6.61}$$

$$function1 = \begin{bmatrix} transfer, & dominating\ object, & the\ cornstalk \\ & acting\ object & \mathrm{DP}_1 \\ & receiving\ object, & u_3 \\ & time, & u_4 \\ & location, & contact\ position \\ & degree, & \begin{array}{c} make\ the\ ear\ stalk \\ arrive\ the\ work\ spot \\ at\ a\ proper\ rate \end{array} \\ & mode, & u_7 \end{bmatrix} \tag{6.62}$$

$$function2 = \begin{bmatrix} exert,\ dominating\ object, & the\ force \\ acting\ object & \mathrm{DP}_1 \\ receiving\ object, & u_3 \\ time, & \begin{matrix} when\ ear\ stalk\ arrives \\ at\ the\ work\ spot \end{matrix} \\ location, & ear\ stalk \\ degree, & \begin{matrix} greater\ than \\ fracture\ stress \end{matrix} \\ mode, & u_7 \end{bmatrix} \tag{6.63}$$

$$function3 = \begin{bmatrix} support,\ dominating\ object, & the\ cornstalks \\ acting\ object & \mathrm{DP}_1 \\ receiving\ object, & u_3 \\ time, & when\ cornstalks\ are\ moving \\ location, & beside\ the\ cornstalks \\ degree, & u_6 \\ mode, & u_7 \end{bmatrix} \tag{6.64}$$

$$\mathrm{DP}_2 = \begin{pmatrix} \begin{matrix} device\ for\ catching \\ and\ gathering, \end{matrix} & function1, & catch\ the\ corn\ ears \\ & function2, & gather\ the\ corn\ ears \\ & configuration, & v_3 \\ & arrangement, & v_4 \end{pmatrix} \tag{6.65}$$

$$function1 = \begin{bmatrix} catch,\ dominating\ object, & the\ corn\ ears \\ acting\ object & \mathrm{DP}_2 \\ receiving\ object, & u_3 \\ time, & when\ the\ corn\ ears\ drop \\ location, & below\ the\ corn\ ears \\ degree, & u_6 \\ mode, & u_7 \end{bmatrix} \tag{6.66}$$

$$function2 = \begin{bmatrix} gather,\ dominating\ object, & the\ corn\ ears \\ acting\ object & \mathrm{DP}_2 \\ receiving\ object, & u_3 \\ time, & \begin{matrix} after\ the\ corn \\ ears\ are\ catched \end{matrix} \\ location, & u_5 \\ degree, & u_6 \\ mode, & u_7 \end{bmatrix} \tag{6.67}$$

Step 4: Document the design functions generated in the mapping process and make sure they are independent of each other. As discussed and seen in Eqs. (6.59) and (6.60), FR_1 and FR_2 are independent of time. And they are implemented by their own design parameter. When DP_1 functions to pick corn ears, it has no effect on FR_2 which is performed at a later time. Similarly, when DP_2 functions to gather fallen corn ears, it does not affect FR_1 which preceded it. This case is referred to as being "independent" according to AD. The design function of this layer can be determined as

$$\begin{pmatrix} FR_1 \\ FR_2 \end{pmatrix} = \begin{bmatrix} X & 0 \\ 0 & X \end{bmatrix} \begin{pmatrix} DP_1 \\ DP_2 \end{pmatrix} \tag{6.68}$$

indicating that the decomposition of this layer results in an uncoupled design.

Step 5: Repeat the above steps till the decomposition of each layer is complete.

The decomposition process is graphically summarized as follows:

The design parameters in Fig. 6.8 are described in conceptual matter-element. The design scheme is not limited to existing components, but is described as conceptual solutions in the form of basic-elements (Eqs. (6.69)–(6.74)).

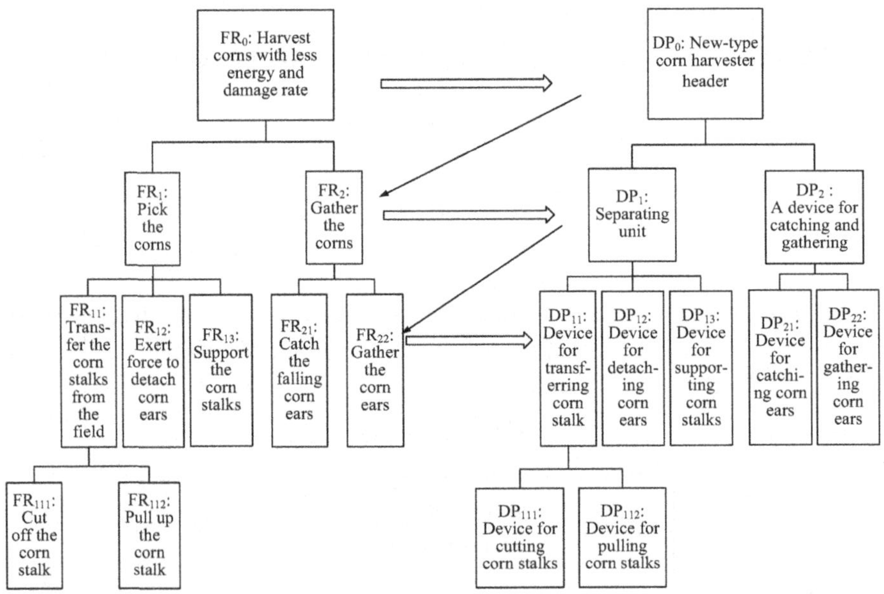

Fig. 6.8 Decomposition process of corn harvester header

$$\mathrm{DP}_{111}=\begin{bmatrix} \begin{matrix} \textit{device for cutting} \\ \textit{corn stalks} \end{matrix} & \textit{function}, & \textit{cutting corn stalks} \\ & \text{configuration}, & \begin{matrix} \textit{device for generating} \\ \textit{the cutting force} \end{matrix} \\ & \textit{arrangement}, & \textit{bottom of the header} \end{bmatrix} \tag{6.69}$$

$$\mathrm{DP}_{112}=\begin{bmatrix} \begin{matrix} \textit{device for pulling} \\ \textit{corn stalks} \end{matrix} & \textit{function}, & \textit{pulling corn stalks} \\ & \textit{configuration}, & \begin{matrix} \textit{device for generating} \\ \textit{the upward force} \end{matrix} \\ & \textit{arrangement}, & \begin{matrix} \textit{above the device} \\ \textit{for cutting corns} \end{matrix} \end{bmatrix} \tag{6.70}$$

$$\mathrm{DP}_{12}=\begin{bmatrix} \begin{matrix} \textit{device for} \\ \textit{detaching corn ears} \end{matrix} & \textit{function}, & \begin{matrix} \textit{exert force to} \\ \textit{detach corn ears} \end{matrix} \\ & \textit{configuration}, & \begin{matrix} \textit{device for generating} \\ \textit{the detaching force} \end{matrix} \\ & \textit{arrangement}, & \begin{matrix} \textit{below the device} \\ \textit{for pulling stalks} \end{matrix} \end{bmatrix} \tag{6.71}$$

$$\mathrm{DP}_{13}=\begin{bmatrix} \begin{matrix} \textit{device for} \\ \textit{supporting cornstalks} \end{matrix} & \textit{function}, & \textit{support the cornstalks} \\ & \textit{configuration}, & \textit{blocking structure} \\ & \textit{arrangement}, & \begin{matrix} \textit{left and right sides} \\ \textit{of cornstalks} \end{matrix} \end{bmatrix} \tag{6.72}$$

$$\mathrm{DP}_{21}=\begin{bmatrix} \begin{matrix} \textit{device for} \\ \textit{catching corn ears} \end{matrix} & \textit{function}, & \textit{catch the falling corn ears} \\ & \textit{configuration}, & \textit{blocking structure} \\ & \textit{arrangement}, & \begin{matrix} \textit{below the device for} \\ \textit{detaching corn ears} \end{matrix} \end{bmatrix} \tag{6.73}$$

$$\mathrm{DP}_{22}=\begin{bmatrix} \begin{matrix} \textit{device for} \\ \textit{gathering corn ears} \end{matrix} & \textit{function}, & \textit{gather thecorn ears} \\ & \textit{configuration}, & \begin{matrix} \textit{device for generating} \\ \textit{the gathering force} \end{matrix} \\ & \textit{arrangement}, & \begin{matrix} \textit{connectedto the} \\ \textit{catchching device} \end{matrix} \end{bmatrix} \tag{6.74}$$

Table 6.6 Divergence of the configuration of each design parameter

DP_{111}: Device for cutting cornstalks	DP_{112}: Device for pulling corn stalks	DP_{12}: Device for detaching corn ears	DP_{13}: Device for supporting corn stalks	DP_{21}: Device for catching corn ears	DP_{22}: Device for gathering corn ears
Reciprocating cutting device, Rotary cutting device	Stalk pulling roller	Picking board	Stalk lifter	Slanting Plate structure, Network structure, Drawer structure	Auger

The configurations conforming to the features can be used as alternative solutions, effectively expanding the solution space. The design scheme is not unique however. More schemes can be developed according to the divergence of the basic-element. For example, DP_{111} can be diverged to multiple types of corn stalks cutting devices.

$$\begin{bmatrix} device\ for & & device\ for \\ cutting & configuration, & generating\ the \\ cornstalks, & & cutting\ force \end{bmatrix}$$
$$-| \begin{bmatrix} device\ for & configuration, & reciprocating \\ cutting\ cornstalks, & & cutting device \\ & & rotary cutting \\ & & device \\ & & \vdots \end{bmatrix} \tag{6.75}$$

Each design parameter can be diverged to multiple solutions, then converged to a few proper ones. Table 6.6 tabulates several feasible solutions below.

Among the extended design schemes, Fig. 6.9 shows an innovative corn harvester header design emerged as a result.

As the corn plant reaches the front of the harvester header, the top of the cornstalk gets into the gap of the stalk pulled rollers, while the bottom of the cornstalk meets the device that cuts corn stalks. Once got cut off, the corn stalk is lifted by the stalk pull-rollers. During this process, the corn ear is lifted together with the cornstalk. When reaching the picking board of the harvester, it is then detached from the cornstalk and falls to the slanting plate, where it is gathered by an auger.

Since there are no effects between each pair of FR_j and DP_i $(i \neq j)$, the design function is established to show that the novel corn harvester header configuration is an uncoupled design.

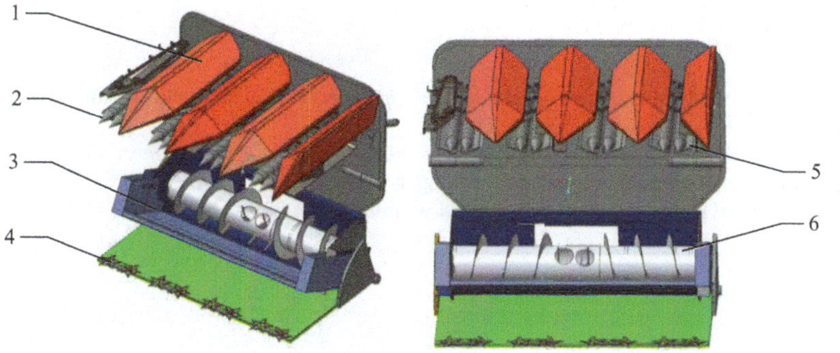

1-DP_{13}: Stalk lifter 2- DP_{112}: Stalk pulling roller 3- DP_{21}: Slanting Plate structure

4- DP_{111}: Rotary cutting device 5- DP_{12}: Picking board 6- DP_{22}: Auger

Fig. 6.9 Novel corn harvester header design

$$\begin{pmatrix} FR_{111} \\ FR_{112} \\ FR_{12} \\ FR_{13} \\ FR_{21} \\ FR_{22} \end{pmatrix} = \begin{bmatrix} X & 0 & 0 & 0 & 0 & 0 \\ 0 & X & 0 & 0 & 0 & 0 \\ 0 & 0 & X & 0 & 0 & 0 \\ 0 & 0 & 0 & X & 0 & 0 \\ 0 & 0 & 0 & 0 & X & 0 \\ 0 & 0 & 0 & 0 & 0 & X \end{bmatrix} \begin{pmatrix} DP_{111} \\ DP_{112} \\ DP_{12} \\ DP_{13} \\ DP_{21} \\ DP_{22} \end{pmatrix} \tag{6.76}$$

6.6 Discussion and Summary

In this chapter, functional requirement is redefined and three criteria are proposed to distinguish FRs from constraints. Moreover, FRs are identified and decomposed from the perspective of inherent sequential logical relationships, which are expressed as a restriction of characteristics in basic-element. Following the proposed criteria and thinking principles, the real functional requirements are identified and well-defined. Corresponding design parameters are determined with a solution procedure for implication function, thus eliminating preconception. Determination of DP is shown to be a process of clarifying the basic-element model sequentially by specifying the characteristics of it. The mapping process is reframed based on the basic-element model which is a solution-based design thinking approach. Extensions of FRs and DPs facilitate the generation of innovative design solutions.

The development of a novel corn harvester header design is elaborated in a step-by-step comprehensive manner to illustrate the implementation of the innovative design methodology. The design solution is shown to be uncoupled and ready to be further expanded. The innovative design methodology can be generalized to significantly enhance design activity and enable design innovation.

References

Cai, W. (1994). *Matter-element model and its application*. Science and Technology Literature Publishing House.

Farid, A. M., & Suh, N. P. (Eds.). (2016). *Axiomatic design in large systems: Complex products, buildings and manufacturing systems*. Springer.

Suh, N. P. (2001). *Axiomatic design: Advances and applications*. New York: Oxford University Press.

Chapter 7
Design Decoupling Methodology

According to the Independence Axiom, each functional requirement should be satisfied with the corresponding design parameter without affecting the others. When the Independence Axiom cannot be satisfied, the coupling problem emerges. Coupling would cause many problems impacting the performance, robustness, reliability, and functionality of the design solution. During the design process, one should try the best to ensure that the solution is uncoupled or decoupled at the least. When a given design solution is a coupled one, the designer must try the best to decouple it. In this chapter, the design decoupling methodology addressing coupling issues is introduced.

7.1 Coupling Identification

The Independence Axiom states that the designer should maintain the independence of the functional requirements when making decision. The design matrices in Eqs. (7.1) and (7.2) are for characterizing product design. Uncoupled design with a diagonal design matrix is ideal, and decoupled design with a triangular design matrix is acceptable, whereas coupled design with other kinds of design matrix is unacceptable.

$$\{\mathrm{FR}\}=[A]\{\mathrm{DP}\} \tag{7.1}$$

$$[A]=\begin{bmatrix} A_{11} & A_{12} & \cdots & A_{1n} \\ A_{21} & A_{22} & \cdots & A_{2n} \\ \vdots & \vdots & & \vdots \\ A_{m1} & A_{m2} & \cdots & A_{mn} \end{bmatrix} \tag{7.2}$$

In Chap. 3, ten specific rules are summarized to help apply the Independence Axiom as follows:

W. Li et al., *Principles of Innovative Design Thinking*,
https://doi.org/10.1007/978-981-19-0485-1_7

Rule 1: "Independence" is not an absolute independence, instead, it is conditional and delimited.

Rule 2: In most cases, uncoupled design cannot be obtained. Instead, the design schemes are usually decoupled design with a triangular matrix.

Rule 3: Decoupling is not necessary for certain acceptable coupling.

Rule 4: The values of the elements in a design matrix usually depend on the designers' consensus.

Rule 5: It is difficult to establish coupling relationship during the concept design process, due to the uncertainty of the specific components.

Rule 6: Designers tend to select a specific solution for each design parameter when applying the AD axioms.

Rule 7: It is better to check the effect of DP on each FR one by one to complete the design matrix.

Rule 8: When there is overlap between FRs, it must be decoupled.

Rule 9: The Independence Axiom can be expressed by mathematical vector graphs. When FRs are independent, DPs are orthogonal to each other, while a DP and its corresponding FR is in a parallel relationship.

Rule 10: Changes should be considered when DPs work to satisfy FRs rather than in the design process.

To help the designers identify coupling, elements A_{ij} in the design matrix are represented in Eq. (7.5) with the FR_i and DP_j being expressed with basic-element models (Eqs. (7.3) and (7.4)).

$$\mathrm{FR}_i = \begin{bmatrix} O_a, & \textit{dominating object}, & u_1 \\ & \textit{acting object} & u_2 \\ & \textit{receiving object}, & u_3 \\ & \textit{time}, & u_4 \\ & \textit{location}, & u_5 \\ & \textit{degree}, & u_6 \\ & \textit{mode}, & u_7 \end{bmatrix} \tag{7.3}$$

$$\mathrm{DP}_i = \begin{bmatrix} O_m, & \textit{function}, & v_1 \\ & \textit{configuration}, & v_2 \\ & \textit{arrangement}, & v_3 \end{bmatrix} \tag{7.4}$$

$$A_{ij}=\frac{\partial \mathrm{FR}_i}{\partial \mathrm{DP}_j}=\frac{\partial \vec{u}}{\partial \vec{v}}=\begin{bmatrix} \frac{\partial u_1}{\partial v_1}, & \frac{\partial u_1}{\partial v_2}, & \frac{\partial u_1}{\partial v_3} \\ \vdots & \frac{\partial u_m}{\partial v_n}, & \vdots \\ \vdots & \vdots & \vdots \\ \frac{\partial u_7}{\partial v_1}, & \frac{\partial u_7}{\partial v_2}, & \frac{\partial u_7}{\partial v_3} \end{bmatrix} \tag{7.5}$$

And the three-steps method for judging the dependence of DP_j and FR_i is proposed in Sect. 5.3.2 as.

Step 1: Establish and clarify all possible state changes of FR_j and Cs.

Step 2: Determine all possible state changes of corresponding DP_j.

Step 3: Judge the effect of DP_j on FR_i $(i \neq j)$ following Eq. (7.5).

An approach for identifying coupling problem is depicted below in a logic flow chart, (Fig. 7.1)

7.2 AD-Based Decoupling Method

Once the coupling problem is identified following the above approach, three theorems from the AD theory are applied to help one identify as to the direction of decoupling, as discussed in Sect. 7.3.

Theorem 4 (Ideal Design): In an ideal design, the number of DPs is equal to the number of FRs and the FRs are always maintained to be independent of each other.

The first decoupling direction is therefore to update the number of DPs to match the number of FRs.

Theorem 8 (Independence and Design Range): A design is an uncoupled design when the designer-specified range is greater than $\left(\sum_{\substack{i\neq j\\ j=1}}^{n} \frac{\partial \mathrm{FR}_i}{\partial \mathrm{DP}_j}\Delta \mathrm{DP}_j\right)$, in which case the non-diagonal elements of the design matrix can be neglected from the design consideration.

The second decoupling direction is therefore to reduce the value of A_{ij}, where $i \neq j$.

Theorem 20 (Design Range and Coupling): If the design ranges of an uncoupled or decoupled design are tightened, they may become coupled designs.

Conversely, if the design ranges of some coupled designs are relaxed, the design may become either uncoupled or decoupled, thus indicating extending the range of design as the third decoupling direction.

The three theorems are followed to guide the decoupling process. However, the practicality of the theorems is limited due to lack of specificity. Consequently, they oftentimes disagree upon the answers, or simply provide none whatsoever, to the following questions: "How to make the number of DPs equal to the number of FRs?", "How to reduce the value of A_{ij}?" and "How to extend the range of design?" There are two fundamental reasons for the poor practicality. The first is that coupled problems are difficult to describe. Secondly, design decoupling is hard to formulate

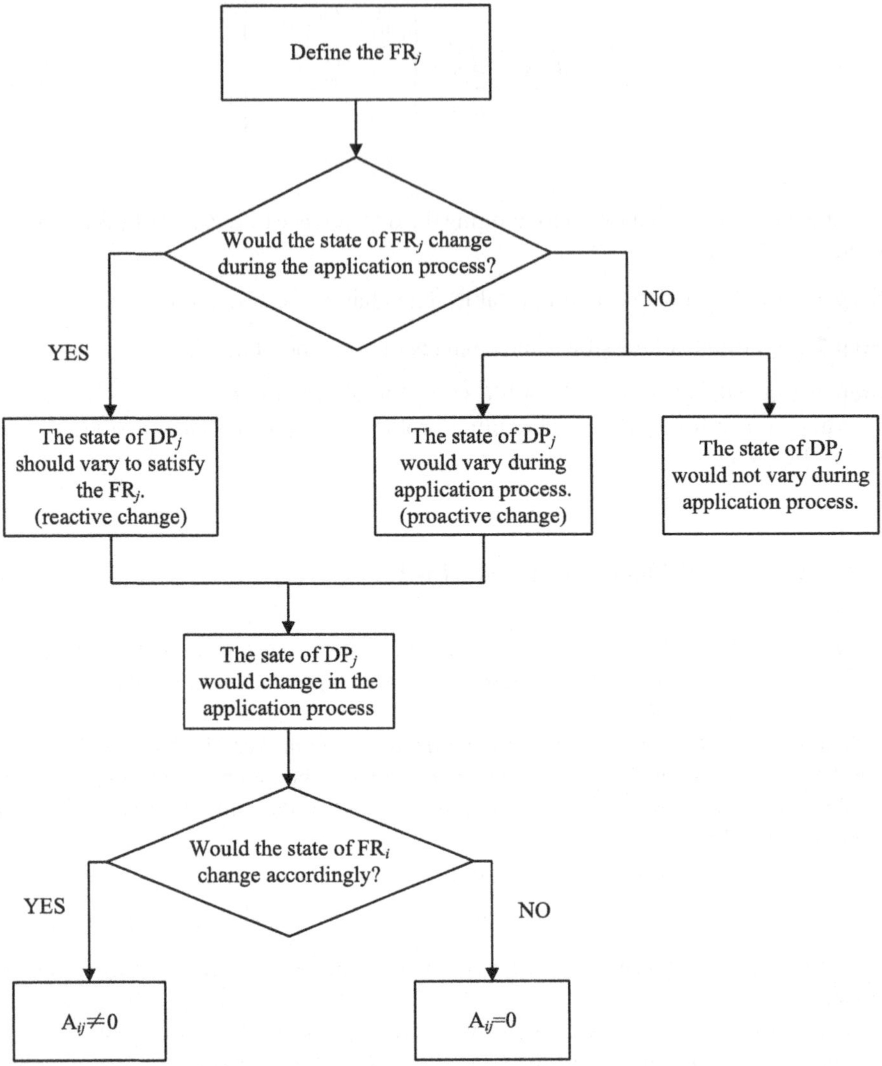

Fig. 7.1 Flow chart for identifying design coupling

when the problem description is not specific or complete. Therefore, it is necessary to comprehend the essences that characterize design coupling and derive the corresponding transformation rules following the theorems to avoid distorted or improper description of the coupled problem.

7.3 General Framework

A design decoupling methodology is proposed addressing the above issues based on the creative synergy of AD and Extenics (Li et al., 2018). The general framework of the design decoupling methodology is established in this section to show how it works in a general case and how to select the proper extension method(s) from Extenics for decoupling.

When a coupling problem is identified, AD is applied to define the coupling problem using a design matrix. The coupling type and the decoupling direction are then determined by the three AD theorems. The coupling and the corresponding maintaining mode model are established in the form of basic-elements (R and M), which provide sets of detailed information for FR_i and DP_j.

$$R=\begin{bmatrix} \textit{coupling}, & \textit{antecedent}, & \mathrm{DP}_j \\ & \textit{consequent}, & \mathrm{FR}_i \\ & \textit{degree}, & A_{ij} \\ & \textit{maintaining mode}, & w_4 \\ & \vdots & \vdots \end{bmatrix} \tag{7.6}$$

$$M=\begin{bmatrix} w_4, & \textit{coupling condition}, & p_1 \\ & \textit{characteristic of antecedent}, & p_2 \\ & \textit{characteristic of consequent}, & p_3 \\ & \textit{impact}, & p_4 \\ & \vdots & \vdots \end{bmatrix} \tag{7.7}$$

The objective of decoupling is to eliminate the impact that DP_j has on FR_i, that is, to reduce the value of the degree in Eq. (7.6) to zero, which is expressed as

$$R'=\begin{bmatrix} \textit{coupling}, & \textit{antecedent}, & \mathrm{DP}_j \\ & \textit{consequent}, & \mathrm{FR}_i \\ & \textit{degree}, & 0 \\ & \textit{maintaining mode}, & w'_4 \\ & \vdots & \vdots \end{bmatrix} \tag{7.8}$$

$$M'=\begin{bmatrix} w'_4, & \textit{coupling condition}, & p'_1 \\ & \textit{characteristic of antecedent}, & p'_2 \\ & \textit{characteristic of consequent}, & p'_3 \\ & \textit{impact}, & p'_4 \\ & \vdots & \vdots \end{bmatrix} \tag{7.9}$$

The characteristics in M includes coupling condition, characteristic of antecedent, characteristic of consequent, impact, and so on. Coupling condition involves circumstance, time, location, and so on. Characteristic of antecedent is related to the basic or intermediate characteristics of DP_j and characteristic of consequent is related to the set of characteristics of FR_i.

The well-defined coupling and maintaining modes (R and M) have the basic-element features of correlation, implication, divergence, opening-up, and conjugation. Extension innovation methods derived from the features are essential to solve the coupling problem.

Firstly, the characteristics of the basic-elements of the coupling problem (R and M) and the basic-elements of the objective (R' and M') are analyzed and the ones crucial to the coupling relationship are identified.

Then, features of the basic-element are selected according to the conditions and objective of the specific coupling problem. With the basic-element R being the description of the coupling, the characteristics in M are given specific focus to solve the coupling problem.

If there exist complex relationships between the entries in M, the Correlation Network Method can be applied to establish the correlations between the basic-elements. If decoupling cannot be reached directly, the Implication System Method can be applied to decompose the target basic-element into several correlative basic-elements. When the present basic-elements M do not solve the coupling problem and more information of M is needed, the divergence analysis can be applied along with the Divergence Tree Method. If M still cannot solve the coupling problem, one can apply the Decomposition-combination Chain Method by exploring the feature of opening-up to develop a new basic-element model of the maintaining mode M using the original ones. Conjugation analysis is generally applied to matter-elements. Since the basic-elements for describing coupling (R, R', M, and M') are usually not typical matter-elements, the conjugation analysis can be used during the transformation transition of the decoupling process.

The proper analysis method considers the condition and objective specific to the case at hand. Extension transformation is applied to formalize and quantify the process of coupling solving. The Extenics set method is applied to describe this process along with the result. The realization of objective R' is regarded as the criterion for judging whether the selected Extenics methods and the result are proper for the coupling problem. Finally, the design function is rebuilt and the new design matrix is evaluated. If the matrix is a diagonal one, the coupling problem is considered solved. Otherwise, it means the solution may lead to a new coupling that is to be resolved by following anew the methodology. In the decoupling process, the expertise of the designer is very important as it dictates the effectiveness and efficiency of problem-solving. The decoupling methodology proposed in this chapter provides a systematic framework for problem-solving while the expertise offers specific information to achieve a concrete scheme. The expertise required for decoupling is decomposed and then refined according to the framework. For example, the required knowledge may be "how to decouple a coupled design?" which is abstract and complex. However, decomposed and refined knowledge may be simply "how to change the condition to make it invalid." Fig. 7.2 shows the general framework of the design decoupling methodology. In the followings, 3 decoupling models are presented, with each being specific to a particular decoupling direction.

7.4 Decoupling Models

7.4.1 Decoupling by Changing Number of DPs

The first decoupling direction is to change the number of DPs to match the number of FRs. In this type of coupling, the number of FRs is greater or less than the number of DPs.

(1) In the first case that the number of FRs is greater than the number of DPs.

For convenience, a representative design is presented as

$$\begin{Bmatrix} FR_1 \\ FR_2 \\ FR_3 \end{Bmatrix} = \begin{bmatrix} A_{11} & 0 \\ 0 & A_{22} \\ 0 & A_{32} \end{bmatrix} \begin{Bmatrix} DP_1 \\ DP_2 \end{Bmatrix} \tag{7.10}$$

where FR_2 and FR_3 share the same design parameter DP_2. To decouple the design, there are several main steps:

Step 1: Determine which FR should be satisfied by the present DP and which should be satisfied by a new DP. Or the two FRs should be satisfied by another two new DPs.

In other words, it is essential to determine which FR should be satisfied by DP_2 and which FR should be satisfied by a new DP. Or, the DP_2 is replaced with two new DPs to satisfy the corresponding FRs. The expertise and knowledge of the designer play a decisive role in this step.

Step 2: Build the model of coupling problem and the objective.

Assume FR_2 should be satisfied by DP_2 and FR_3 should be satisfied by a new DP. Then the relationship of coupling is between DP_2 and FR_3. The relation-element model as well as the maintaining mode model are, respectively,

$$R = \begin{bmatrix} coupling, & antecedent, & DP_2 \\ & consequent, & FR_3 \\ & degree, & A_{32} \\ & maintaining\ mode, & w_4 \\ & \vdots & \vdots \end{bmatrix} \tag{7.11}$$

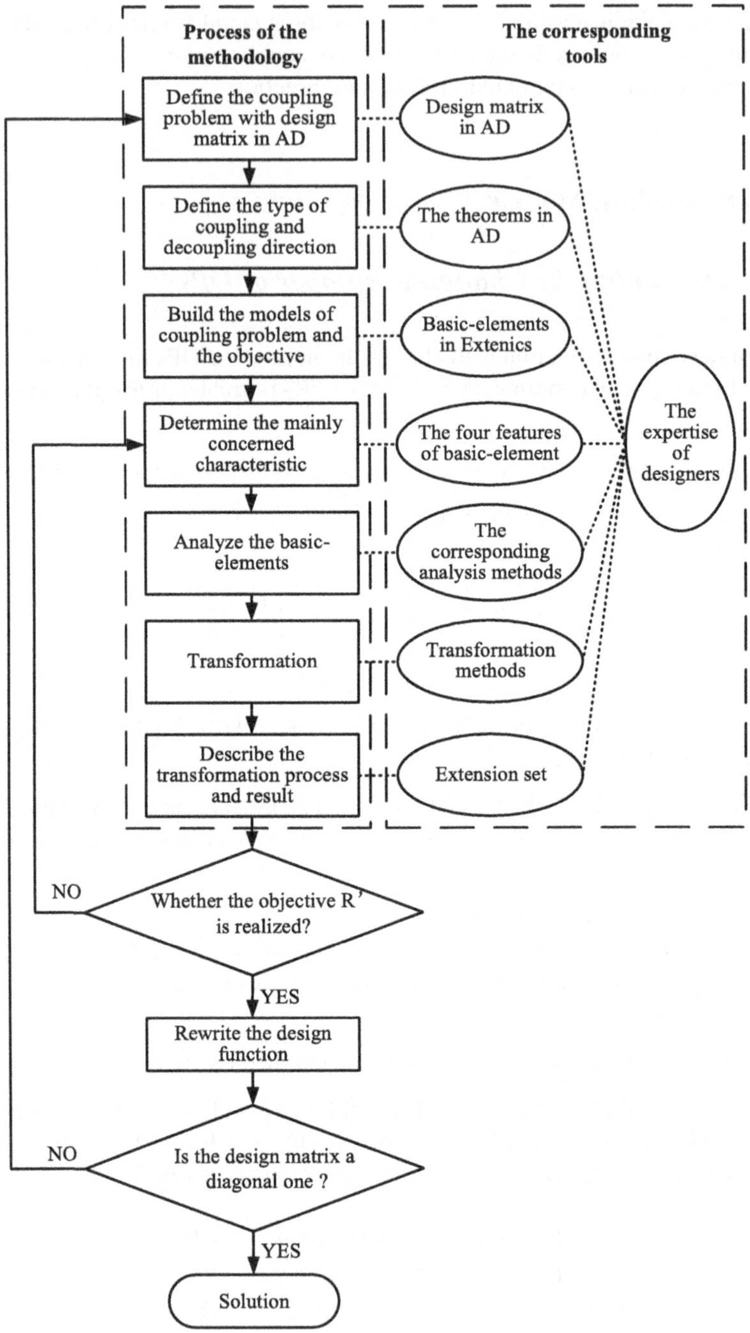

Fig. 7.2 General framework of decoupling methodology

$$M = \begin{bmatrix} w_4, & coupling\ condition, & p_1 \\ & characteristic\ of\ antecedent, & p_2 \\ & characteristic\ of\ consequent, & p_3 \\ & impact, & p_4 \\ & \vdots & \vdots \end{bmatrix} \tag{7.12}$$

The parameters $(p_1, p_2, \ldots)$ in Eq. (7.12) can be determined with the basic-element models of FR_3 (Eq. 7.13) and DP_2 (Eq. 7.14).

$$\mathrm{FR}_3 = \begin{bmatrix} O_a, & dominating\ object, & u_1 \\ & acting\ object & \mathrm{DP}_2 \\ & receiving\ object, & u_3 \\ & time, & u_4 \\ & location, & u_5 \\ & degree, & u_6 \\ & mode, & u_7 \end{bmatrix} \tag{7.13}$$

$$\mathrm{DP}_2 = \begin{bmatrix} O_m, & function, & v_1 \\ & configuration, & v_2 \\ & arrangement, & v_3 \end{bmatrix} \tag{7.14}$$

The objective of this kind of decoupling is to transform the 3 × 2 matrix in Eq. (7.10) to a 3 × 3 diagonal matrix. The ideal design function would look like Eq. (7.15).

$$\begin{Bmatrix} \mathrm{FR}_1 \\ \mathrm{FR}_2 \\ \mathrm{FR}_3 \end{Bmatrix} = \begin{bmatrix} A_{11} & 0 & 0 \\ 0 & A_{22} & 0 \\ 0 & 0 & A_{33} \end{bmatrix} \begin{Bmatrix} \mathrm{DP}_1 \\ \mathrm{DP}_2 \\ \mathrm{DP}_3 \end{Bmatrix} \tag{7.15}$$

To achieve the ideal design function, the relationship between DP_2 and FR_3 should be removed firstly. Then a new design parameter should be introduced as DP_3 to satisfy FR_3. Consequently, the objective models are

$$R' = \begin{bmatrix} coupling, & antecedent, & \mathrm{DP}_2 \\ & consequent, & \mathrm{FR}_3 \\ & degree, & 0 \\ & maintaining\ mode, & w'_4 \\ & \vdots & \vdots \end{bmatrix} \tag{7.16}$$

$$M' = \begin{bmatrix} w'_4, & coupling\ condition, & p'_1 \\ & characteristic\ of\ antecedent, & p'_2 \\ & characteristic\ of\ consequent, & p'_3 \\ & impact, & p'_4 \\ & \vdots & \vdots \end{bmatrix} \tag{7.17}$$

$$\mathrm{FR}_3 = \begin{bmatrix} O_a, & dominating\ object, & u_1 \\ & acting\ object & \mathrm{DP}_3 \\ & receiving\ object, & u_3 \\ & time, & u_4 \\ & location, & u_5 \\ & degree, & u_6 \\ & mode, & u_7 \end{bmatrix} \tag{7.18}$$

Step 3: Make A_{32} equal to zero through transformation.

Per AD, the goal is to reduce A_{32} to zero. The knowledge needed is therefore how to reduce A_{32}. Per the design decoupling methodology, making A_{32} equal to zero is to transform the coupling model R to R'. Thus, the knowledge needed here is how to change the value form w_4 to w'_4, which are expressed with M and M'. Consequently, the coupling problem can be solved with the transformation of M to M'. The value p_4 in M represents the specific impact of the coupling, the ideal state of which is "no effect". As discussed in Sect. 4.2, M can be analyzed by exploring the features of correlation, implication, divergence, opening-up and conjugation. The Correlation Network Method can be applied to study the relationship between p_4 and other factors in M. If the target basic-elements R' and M' cannot be reached directly, it can be decomposed into several correlative basic-elements according to the Implication System Method. The Divergence Analysis Method can be applied to exploit more information to reduce A_{32}. And the Decomposition-combination Chain Method can help develop new models of M to find a way to reduce A_{32}. The detail steps of the method have been presented in previous part of the book.

A proper analysis method is to be selected by considering the specific condition and objective as required and the primary characteristic of M is to be determined so as to facilitate the reduction of A_{32}. Transformation can then be conducted to solve the design problem. The design decoupling methodology provides an analytical framework within which the expertise and knowledge of a designer are applied in an organized manner.

Step 4: Introduce DP_3 to satisfy FR_3.

FR_3 is analyzed with affair-element shown in Eq. (7.13) to introduce the new design parameter DP_3 which will be represented as a matter-element model with the basic characteristics of function, configuration and arrangement. The expertise and knowledge help designers extract the effective information of the affair-element model of

FR_3, while Extenics helps apply the information to determine the basic characteristics in DP_3. Firstly, the function of DP_3 will be determined with FR_3. To realize the function, the implication analysis can be applied to search for inferior basic-elements of DP_3, while the divergence analysis can help exploit more information of DP_3 for basic characteristics. Moreover, the relations between the basic characteristics in DP_3 should be considered to determine the characteristics of DP_3, where the Correlation Analysis Method can be applied. And the Decomposition-combination Chain Method as well as the conjugation analysis will help develop new DP_3 to satisfy FR_3. The application conditions and goals as well as detail steps of the methods were discussed in Chap. 4. The designers can select the proper method according to the practical situation as well as their expertise. In this way, the effect that DP_2 initially had on FR_3 is now transferred to DP_3.

Step 5: Rewrite the design function and check the design matrix.

When the design matrix is a diagonal matrix, the decoupling is complete. If not, one is to start over from the first step.

(2) In the other type of coupling, the number of FRs is less than the number of DPs. The corresponding design function is

$$\begin{Bmatrix} FR_1 \\ FR_2 \end{Bmatrix} = \begin{bmatrix} A_{11} & 0 & A_{13} \\ 0 & A_{22} & A_{23} \end{bmatrix} \begin{Bmatrix} DP_1 \\ DP_2 \\ DP_3 \end{Bmatrix} \tag{7.19}$$

This kind of coupling is called a redundant design. To turn the design described by Eq. (7.19) into an ideal uncoupled design defined in Eq. (7.15), two approaches are available from AD: by fixing DP_3 so that it does not act as a design parameter or by making the coefficients associated with DP_3 (A_{13} and A_{23}) equal to zero. If the design matrix is different from the one shown above, other appropriate design elements should be made zero or other appropriate DPs should be fixed to reduce the variance of the FRs (Suh, 2001). Fixing the appropriate DPs can be done through a simple operation depending on the designer's expertise, which is not discussed herein.

7.4.2 Decoupling by Reducing Value of A_{ij}

The second decoupling direction is to reduce the value of A_{ij}, i.e. to reduce the impact of DP_j on FR_i. A representative design is presented in Eq. (7.20). The coupling between FR_1 and DP_3 is discussed to illustrate the decoupling process.

$$\begin{Bmatrix} FR_1 \\ FR_2 \\ FR_3 \end{Bmatrix} = \begin{bmatrix} A_{11} & 0 & A_{13} \\ A_{21} & A_{22} & 0 \\ A_{31} & A_{32} & A_{33} \end{bmatrix} \begin{Bmatrix} DP_1 \\ DP_2 \\ DP_3 \end{Bmatrix} \tag{7.20}$$

Step 1: Build the model of coupling problem and define the objective

The corresponding relation-element model and the maintaining mode model are established, respectively,

$$R = \begin{bmatrix} coupling, & antecedent, & \mathrm{DP}_3 \\ & consequent, & \mathrm{FR}_1 \\ & degree, & A_{13} \\ & maintaining\ mode, & w_4 \\ & \vdots & \vdots \end{bmatrix} \tag{7.21}$$

$$M = \begin{bmatrix} w_4, & coupling\ condition, & p_1 \\ & characteristic\ of\ antecedent, & p_2 \\ & characteristic\ of\ consequent, & p_3 \\ & impact, & p_4 \\ & \vdots & \vdots \end{bmatrix} \tag{7.22}$$

$$R' = \begin{bmatrix} coupling, & antecedent, & \mathrm{DP}_3 \\ & consequent, & \mathrm{FR}_1 \\ & degree, & 0 \\ & maintaining\ mode, & w'_4 \\ & \vdots & \vdots \end{bmatrix} \tag{7.23}$$

$$M' = \begin{bmatrix} w'_4, & couplingcondition, & p'_1 \\ & characteristic\ of\ antecedent, & p'_2 \\ & characteristic\ of\ consequent, & p'_3 \\ & impact, & p'_4 \\ & \vdots & \vdots \end{bmatrix} \tag{7.24}$$

Step 2: Determine the mainly concerned characteristics

In this kind of coupling, FR_1 is satisfied by the combined action of DP_1 and DP_3. Decoupling is to reduce the value of A_{13}, which is similar to Step 3 in Sect. 7.4.1. DP_3 has an impact on FR_1 while being the primary parameter of FR_3. The relationship between DPs and FRs is so complex that it complicates the characteristics that are in the maintaining mode model (M) of the coupling. To achieve the objective R', the coupling conditions or the design parameter DP_3 needs be changed. To determine which characteristic can be changed for problem-solving, the features of basic-element as well as expertise are considered. The features including correlation, implication, divergence, opening-up and conjugation provide the thinking orientation while expertise is applied to specific decision.

Step 3: Analyze and transform the basic-elements

Extenics analysis methods can be applied for the transformation to reduce the value of A_{ij}. Then the solution to the coupling problem can be obtained.

The other couplings represented by A_{21}, A_{31}, and A_{32} can be decoupled in the same way or be accepted as a decoupled design.

Step 4: Validate the result

Firstly, the coupling model is rebuilt with the solution to see whether the objective R' is realized. If it is achieved, the design function is rewritten and the design matrix is confirmed. When the design matrix is a diagonal or triangle matrix, the decoupling is complete. If not, one is to start over from the first step according to Fig. 7.2.

7.4.3 Decoupling by Extending Design Range

The third decoupling direction is to extend the range of design. As the range of design is usually settled in a reasonable range at the beginning of the design effort, this kind of decoupling may seem deceivingly easy.

Assume that the design function is of the same form as Eq. and a coupling exists between DP_1 and FR_2. Compared to the decoupling method in Sect. 7.4.2, the third decoupling direction is to change the range of design instead of the design parameter. It is suitable for instances where changing the design parameter is inadvisable. AD indicates that, if the design range of a coupled design is relaxed, the design may become either uncoupled or decoupled. The problems, including how to determine the design range that needs be extended and how to extend the range, can be addressed following the decoupling framework below:

Step 1: Build the model of coupling problem and define the objective

Since the design function is of the same form as Eq. (7.20), the relative relation-element model and maintaining mode model can be formulated by following Eqs. (7.21)–(7.23).

Step 2: Determine primary characteristics of concern

To determine which characteristics are associated with the design range that leads to the coupling, the features of basic-element are explored.

Step 3: Analyze and transform the basic-elements

Proper Extenics analysis methods and transformation methods are applied to determine whether the design range can be extended along with how to extend the design range to resolve the coupling problem.

Step 4: Validate the result

The coupling model is rebuilt with the solution to see whether the coupling is resolved. If yes, the design function and design matrix are updated.

7.5 Illustrative Examples

Example 7.1 The corn harvester header is considered as the design object. At present, the most widely used type is the snapping plate-stalk pulling roller combined picker shown in Fig. 7.3.

When the stalk-pulling rollers roll, the corn stalks are pulled down till the ears reach the snapping plate, where the corn ears are picked. Then it's necessary to keep the removed corn ears from falling. The design function is

$$\begin{Bmatrix} \mathrm{FR}_1 \\ \mathrm{FR}_2 \\ \mathrm{FR}_3 \end{Bmatrix} = \begin{bmatrix} A_{11} & 0 \\ 0 & A_{22} \\ 0 & A_{32} \end{bmatrix} \begin{Bmatrix} \mathrm{DP}_1 \\ \mathrm{DP}_2 \end{Bmatrix} \tag{7.25}$$

where

FR_1 = Pull stalks.

$$\mathrm{FR}_1 = \begin{bmatrix} \textit{pull, dominating object,} & \textit{stalks} \\ \textit{actingobject} & \mathrm{DP}_1 \\ \textit{receiving object,} & u_3 \\ \textit{time,} & u_4 \\ \textit{location,} & u_5 \\ \textit{degree,} & \textit{at a proper speed} \\ \textit{mode,} & u_7 \end{bmatrix} \tag{7.26}$$

FR_2 = Catch removed corn ears.

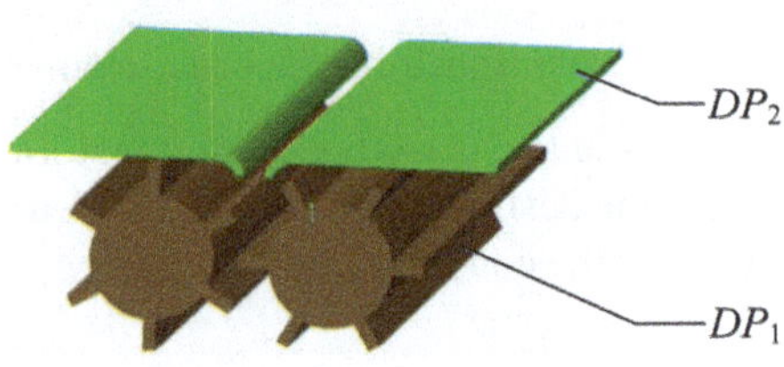

Fig. 7.3 Snapping plate-stalk pulling roller combined picker

DP_1- Stalk pulling rollers DP_2- Snapping plates

$$\mathrm{FR}_2 = \begin{bmatrix} \textit{catch, dominating object,} & \textit{the removed corn ears} \\ \textit{actingobject} & \mathrm{DP}_2 \\ \textit{receiving object,} & u_3 \\ \textit{time,} & \textit{afterears were removed} \\ \textit{location,} & u_5 \\ \textit{degree,} & u_6 \\ \textit{mode,} & u_7 \end{bmatrix} \quad (7.27)$$

$\mathrm{FR}_3 =$ Sever corn ears from stalks.

$$\mathrm{FR}_3 = \begin{bmatrix} \textit{sever, dominating object,} & \textit{corn ears} \\ \textit{actingobject} & \mathrm{DP}_2 \\ \textit{receiving object,} & \textit{whole}-\textit{plant corn} \\ \textit{time,} & \textit{when ears arrive at the plates} \\ \textit{location,} & \textit{edge of the snapping plates} \\ \textit{degree,} & \textit{completely detached from stalks} \\ \textit{mode,} & \begin{matrix}\textit{break the connection} \\ \textit{between ear and stalk}\end{matrix} \end{bmatrix} \quad (7.28)$$

$\mathrm{DP}_1 =$ Stalk-pulling rollers

$$\mathrm{DP}_1 = \begin{bmatrix} \textit{Stalk}-\textit{pulling rollers,} & \textit{function,} & \textit{Pull stalks} \\ & \textit{configuration,} & \textit{roller} \\ & \textit{arrangement,} & v_3 \end{bmatrix} \quad (7.29)$$

$\mathrm{DP}_2 =$ Snapping plates

$$\mathrm{DP}_2 = \begin{bmatrix} \textit{Snapping plates,} & \textit{function,} & \begin{matrix}\textit{Catch removed corn ears} \\ \textit{andsever corn ears from stalks}\end{matrix} \\ & \textit{configuration,} & \textit{plate} \\ & \textit{arrangement,} & v_3 \end{bmatrix} \quad (7.30)$$

It is evident from the design function that this is a coupled design where the numbers of DPs are less than the number of FRs. The snapping plate has the functions of both severing the corn ears from stalks and catching the removed corn ears. Obviously, the decoupling in this example corresponds to the first direction, which is discussed in Sect. 7.4.1. Per AD, the design may be decoupled by the addition of a new DP. The problem to be solved is "how to find an uncoupled solution which satisfies this set of FRs with the addition of a new DP"—which can be very confusing. The design decoupling methodology proposed above is applied to illustrate the flow of problem-solving.

Step 1: Determine which FR is to be satisfied by the present DP and which by a new DP

Catching removed corn ears relies on a surface to support the ears while severing corn ears from stalks is realized with a force large enough to fracture. Snapping plates provide the surface needed by FR_2. And the force can be applied in various ways in addition to what is facilitated by the snapping plates. Therefore, DP_2 is applied to satisfy FR_2 while a new design parameter DP_3 is introduced for FR_3.

Step 2: Build the model of the coupling problem and identify the objective

The coupling is between DP_2 and FR_3. The basic-element models of FR_3 and DP_2 are established for description.

$$FR_3 = \begin{bmatrix} \textit{sever}, \textit{dominating object}, & \textit{corn ears} \\ \textit{actingobject} & DP_2 \\ \textit{receiving object}, & \textit{whole} - \textit{plant corn} \\ \textit{time}, & \textit{when ears arrive at the plates} \\ \textit{location}, & \textit{edge of the snapping plates} \\ \textit{degree}, & \textit{completely detached} \\ \textit{mode}, & \begin{matrix} \textit{break the connection} \\ \textit{between ear and stalk} \end{matrix} \end{bmatrix} \tag{7.31}$$

$$DP_2 = \begin{bmatrix} \textit{snapping plates}, & \textit{function}, & \begin{matrix} \textit{sever corn ears from stalks} \\ \textit{and catch removed corn ears} \end{matrix} \\ & \textit{configuration}, & \textit{snapping plate} \\ & \textit{arrangement}, & \begin{matrix} \textit{symmetrically arranged over} \\ \textit{the rollers with a proper gap} \end{matrix} \end{bmatrix} \tag{7.32}$$

The relation-element model and the maintaining mode model are, respectively,

$$R = \begin{bmatrix} \textit{coupling}, & \textit{antecedent}, & \textit{snapping plates} \\ & \textit{consequent}, & \textit{sever corn ears} \\ & \textit{degree}, & A_{32} \\ & \textit{maintainingmode}, & w_4 \end{bmatrix} \tag{7.33}$$

$$M = \begin{bmatrix} w_4, & \textit{coupling condition1}, & \textit{the contact between ears and the plate} \\ & \textit{coupling condition2}, & \begin{matrix} \textit{the collision force is} \\ \textit{greater than fracture stress} \end{matrix} \\ & \textit{characteristic of antecedent}, & \begin{matrix} \textit{the reaction to ears} \\ \textit{provided by the plate} \end{matrix} \\ & \textit{characteristic of consequent}, & \begin{matrix} \textit{break the connection} \\ \textit{between ears and stalk} \end{matrix} \\ & \textit{impact}, & \textit{break the ears totally} \end{bmatrix} \tag{7.34}$$

As discussed, the relationship between DP_2 and FR_3 is to be removed. A new design parameter is introduced as DP_3 to satisfy FR_3. The objective models are

$$R' = \begin{bmatrix} coupling, & antecedent, & snapping\ plate \\ & consequent, & sever\ corn\ ears \\ & degree, & 0 \\ & maintaining mode, & w'_4 \end{bmatrix} \tag{7.35}$$

$$FR_3 = \begin{bmatrix} sever, dominating\ object, & corn\ ear \\ acting object & DP_3 \\ receiving\ object, & whole - plant\ corn \\ time, & u_4 \\ location, & u_5 \\ degree, & completely\ detached \\ mode, & u_7 \end{bmatrix} \tag{7.36}$$

Step 3: Make A_{32} equal to zero through transformation

As discussed above, making A_{32} equal to zero is the same as transforming the coupling model R to R', which can be achieved with the change of Condition 1 in Eq. (7.34).

$$\begin{aligned} &(w_4,\ coupling\ condition\ 1,\ the\ contact\ between\ ears\ and\ the\ plate) \\ &-|(w'_4,\ coupling\ condition\ 1,\ no - contact\ between\ ears\ and\ the\ plate) \end{aligned} \tag{7.37}$$

That the ears do not engage the plate helps avoid invading the ears, thus avoiding the coupling between DP_2 and FR_3, thus R' is achieved.

Step 4: Introduce DP_3 to satisfy FR_3.

Rewrite the affair-element of FR_3 since the corresponding design parameter has changed.

$$FR_3 = \begin{bmatrix} Sever, dominating\ object, & corn\ ears \\ acting object & DP_3 \\ receiving\ object, & whole - plant\ corn \\ time, & u_4 \\ location, & u_5 \\ degree, & u_6 \\ mode, & \begin{matrix} no - contact between \\ ears and the plate \end{matrix} \end{bmatrix} \tag{7.38}$$

DP_3 should provide the force to sever corn ears while preventing the ears from contacting the plate. A new DP_3 with cutting function is selected herein as

$$\mathrm{DP}_3 = \begin{bmatrix} \mathrm{DP}_3, & \textit{function}, & \textit{severcornears} \\ & \textit{configuration}, & \begin{matrix} \textit{bladesenveloped} \\ \textit{withrubbersleeves} \end{matrix} \\ & \textit{arrangement}, & \begin{matrix} \textit{symmetricallyarranged} \\ \textit{onthesnappingplates} \end{matrix} \end{bmatrix} \tag{7.39}$$

$$v_1 = \begin{bmatrix} \textit{Sever}, \textit{dominating object}, & \textit{corn ears} \\ \textit{actingobject} & \mathrm{DP}_3 \\ \textit{receiving object}, & \textit{whole} - \textit{plant corn} \\ \textit{time}, & u_4 \\ \textit{location}, & u_5 \\ \textit{degree}, & u_6 \\ \textit{mode}, & \begin{matrix} \textit{cut the ear stalk with} \\ \textit{no} - \textit{contactbetween} \\ \textit{earsandtheplate} \end{matrix} \end{bmatrix} \tag{7.40}$$

In this case, FR_3 is satisfied. The new design is shown in Fig. 7.4.

Step 5: Rewrite the design function and update the design matrix

The design function is updated to a 3×3 diagonal matrix. The new DP_2 is transformed to DP'_2 as shown in Eq. (7.41) while DP_3 is a cutting structure as shown in Eq. (7.39).

$$\mathrm{DP}'_2 = \begin{bmatrix} \textit{snapping plates}, & \textit{function}, & \textit{hold the ears} \\ & \textit{configuration}, & \textit{snappingplate} \\ & \textit{arrangement}, & \begin{matrix} \textit{symmetrically arranged over} \\ \textit{the rollers with a proper gap} \end{matrix} \end{bmatrix} \tag{7.41}$$

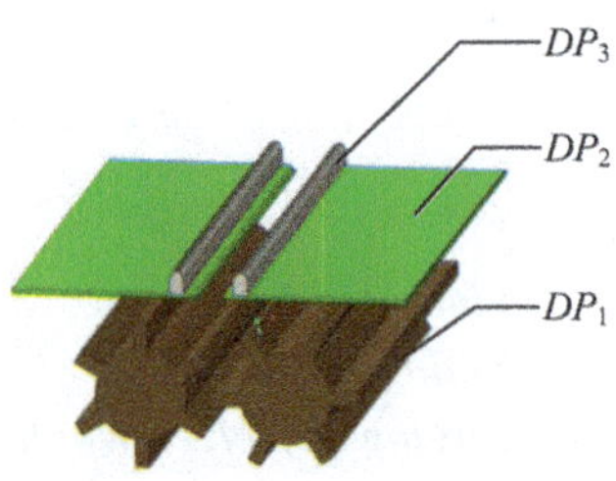

DP1- Stalk pulling rollers DP2- Snapping plates DP3- Cutting structure

Fig. 7.4 New design of corn harvester header

As discussed above, the coupling between DP_2 and FR_3 is eliminated without inducing new coupling issues, and the design function is updated to a 3 × 3 diagonal matrix of the same form as Eq. (7.15). Thus, the coupling problem is solved.

Example 7.2 The refrigerator door example found in Page 20 in Ref. 1 (Suh, 2001) is a typical case of a decoupled design. The example is considered herein to illustrate the application of the introduced decoupling method.

The required FRs of the door are redefined as.
FR_1 = Provide access to the items stored in the refrigerator

$$FR_1 = \begin{bmatrix} provide, \; dominating\ object, & access \\ actingobject & DP_1 \\ receiving\ object, & u_3 \\ time, & u_4 \\ location, & u_5 \\ degree, & conveniently \\ mode, & \end{bmatrix} \tag{7.42}$$

FR_2 = Minimize energy loss

$$FR_2 = \begin{bmatrix} mininmize, \; dominating\ object, & engergy\ loss \\ actingobject & DP_2 \\ receiving\ object, & u_3 \\ time, & u_4 \\ location, & u_5 \\ degree, & u_6 \\ mode, & u_7 \end{bmatrix} \tag{7.43}$$

DP_1 = Vertically hung door

$$DP_1 = \begin{bmatrix} \begin{matrix} Vertically \\ hung\ door, \end{matrix} & function, & \begin{matrix} Provide\ access\ to\ the\ items \\ stored\ in\ the\ refrigerator \end{matrix} \\ & configuration, & door \\ & arrangement, & Vertically\ hung \end{bmatrix} \tag{7.44}$$

DP_2 = Thermal insulation material in the door

$$DP_2 = \begin{bmatrix} \begin{matrix} Thermal\ insulation \\ material, \end{matrix} & function, & Minimize\ energy\ loss \\ & configuration, & v_2 \\ & arrangement, & v_3 \end{bmatrix} \tag{7.45}$$

The corresponding design matrix is

$$\begin{Bmatrix} \mathrm{FR}_1 \\ \mathrm{FR}_2 \end{Bmatrix} = \begin{bmatrix} A_{11} & 0 \\ A_{21} & A_{22} \end{bmatrix} \begin{Bmatrix} \mathrm{DP}_1 \\ \mathrm{DP}_2 \end{Bmatrix} \tag{7.46}$$

Step 1: Build the model of the coupling problem and define the objective

$$R = \begin{bmatrix} coupling, & antecedent, & vertically\ hung\ door \\ & consequent, & minimizeenergyloss \\ & degree, & A_{21} \\ & maintainingmode, & v_{r4} \\ & \vdots & \vdots \end{bmatrix} \tag{7.47}$$

$$M = \begin{bmatrix} v_{r4}, & couplingcondition, & whenthedooropens \\ & characteristicofantecedent, & absenceof\ verticalobstruct \\ & characteristicofconsequent, & cold\ air\ is\ heavier \\ & impact, & \begin{matrix} cold\ air\ flows\ down\ with \\ a\ high\ speed\ of\ flowing\ down \end{matrix} \end{bmatrix} \tag{7.48}$$

$$R' = \begin{bmatrix} coupling, & antecedent, & DP_1 \\ & consequent, & FR_2 \\ & degree, & 0 \\ & maintainingmode, & v'_{r4} \\ & \vdots & \vdots \end{bmatrix} \tag{7.49}$$

Step 2: Determine the primary characteristics of concern

The corresponding correlation network of the factors in M is given in Fig. 7.5,

To change the coupling degree (p_5: high speed of flowing down) to zero, one can change the characteristic of antecedent (p_2: absence of vertical obstruct).

Step 3: Analyze and transform the basic-elements

In this case, divergence is more evident among the features of p_2 (absence of vertical obstruct). Transformation following the divergence analysis is performed.

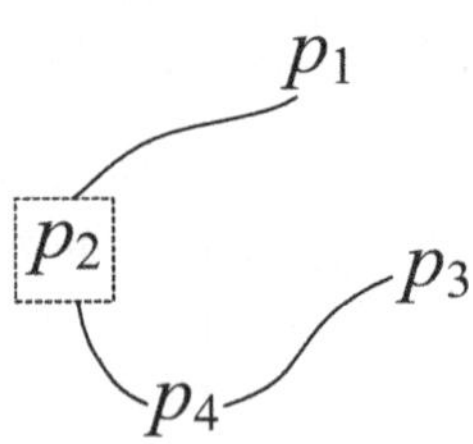

Fig. 7.5 The correlation network of factors in Example 7.2

$$\begin{aligned}&(characteristic\ of\ antecedent,\ absence\ of\ vertical\ obstruct)\\&-|(characteristic\ of\ antecedent,\ absence\ of\ horizontal\ obstruct)\end{aligned} \quad (7.50)$$

Step 4: Validate the result

The characteristic of antecedent (DP_1) is altered according to the transformation, inducing a change of DP_1 to a new design parameter—a horizontally hung door as a result. The objective R' is therefore achieved. Consequently, the design function can be updated to be

$$\begin{Bmatrix} FR_1 \\ FR_2 \end{Bmatrix} = \begin{bmatrix} A_{11} & 0 \\ 0 & A_{22} \end{bmatrix} \begin{Bmatrix} DP_1 \\ DP_2 \end{Bmatrix} \quad (7.51)$$

which is identical to the decoupled design found in Ref. 1.

Example 7.3 Keeping structural integrity and providing mounting points are the 2 primary functions of the chassis of a harvester. The design function is.

$$\begin{Bmatrix} FR_1 \\ FR_2 \end{Bmatrix} = \begin{bmatrix} A_{11} & 0 \\ A_{21} & A_{22} \end{bmatrix} \begin{Bmatrix} DP_1 \\ DP_2 \end{Bmatrix} \quad (7.52)$$

FR_1 = Provide mounting points

$$FR_1 = \begin{bmatrix} \multicolumn{2}{c}{provide,\ dominating\ object,\ mountingpoints} \\ actingobject & DP_1 \\ receiving\ object, & u_3 \\ time, & u_4 \\ location, & on\ the\ chassis \\ degree, & accurately \\ mode, & u_7 \end{bmatrix} \quad (7.53)$$

FR_2 = Keep structural integrity of the chassis

$$FR_2 = \begin{bmatrix} \multicolumn{2}{c}{keep,\ dominating\ object,\ structuralintegrity} \\ actingobject & DP_1 \\ receiving\ object, & chassis \\ time, & u_4 \\ location, & u_5 \\ degree, & firmly \\ mode, & u_7 \end{bmatrix} \quad (7.54)$$

DP_1 = Configuration of the chassis

$$DP_1 = \begin{bmatrix} \begin{matrix} \textit{Configuration} \\ \textit{of the chassis,} \end{matrix} & \textit{function,} & \textit{Provide mounting points} \\ & \textit{configuration,} & v_2 \\ & \textit{arrangement,} & \textit{bottom oftheharvester} \end{bmatrix} \quad (7.55)$$

DP_2 = Material property of the chassis

$$DP_2 = \begin{bmatrix} \begin{matrix} \textit{Material property} \\ \textit{of the chassis,} \end{matrix} & \textit{function,} & \begin{matrix} \textit{Keep structural integrity} \\ \textit{of the chassis} \end{matrix} \\ & \textit{configuration,} & v_2 \\ & \textit{arrangement,} & v_3 \end{bmatrix} \quad (7.56)$$

This is a difficult design problem given the insufficient information on the coupling. Apply the decoupling methodology to the example problem as follows:

Step 1: Build the model of the coupling problem with the objective identified.

$$R = \begin{bmatrix} \textit{coupling,} & \textit{antecedent,} & \textit{configuration of the chassis} \\ & \textit{consequent,} & \textit{ensure structural integrity} \\ & \textit{degree,} & A_{21} \\ & \textit{maintainingmode,} & w_4 \\ & \vdots & \vdots \end{bmatrix} \quad (7.57)$$

$$M = \begin{bmatrix} w_4, & \textit{couplingcondition,} & \begin{matrix} \textit{the loads to sustain are close to} \\ \textit{the ultimate strength of the chassis} \end{matrix} \\ & \begin{matrix} \textit{characteristic} \\ \textit{ofantecedent,} \end{matrix} & \begin{matrix} \textit{stressconcentration} \\ \textit{causedbythestructures} \end{matrix} \\ & \begin{matrix} \textit{characteristic} \\ \textit{ofconsequent,} \end{matrix} & \textit{strength} \\ & \textit{impact,} & \begin{matrix} \textit{weakenthestrength} \\ \textit{to fracture the structure} \end{matrix} \end{bmatrix} \quad (7.58)$$

$$R' = \begin{bmatrix} \textit{coupling,} & \textit{antecedent,} & \textit{configuration of the chassis} \\ & \textit{consequent,} & \textit{ensure structural integrity} \\ & \textit{degree,} & 0 \\ & \textit{maintainingmode,} & w'_4 \\ & \vdots & \vdots \end{bmatrix} \quad (7.59)$$

Step 2: Determine the primary characteristics of concern.

In this case, the complex relationship can be expressed with the correlation network of the factors in M as (Fig. 7.6).

Obviously, the coupling condition (the loads to sustain are close to the ultimate strength of the chassis) is related to the design range.

Step 3: Analyze and transform the basic-elements.

According to opening-up analysis method, any basic-element can be decomposed into several basic-elements under certain condition, i.e.

$$(weight\ to\ bear,\ g)//\{(weight\ of\ grains, g_1), (weight\ of\ the\ machine, g_2)\} \quad (7.60)$$

Since the design range is the strength of the frame beyond the strength caused by the heaviest weight of the harvester, to extend the design range is to reduce the payload, as shown in Fig. 7.7, where A and B represent the strength caused by the heaviest weight of the harvester, design range is the interval greater than A or B.

As the weight of the harvested grains can be separated by a carrier vehicle, the payload can be reduced.

Step 4: Validate the result.

Once the carrier vehicle is introduced, the material property of the chassis (DP_2) can be manipulated to ensure the structural integrity of the chassis (FR_2) and not

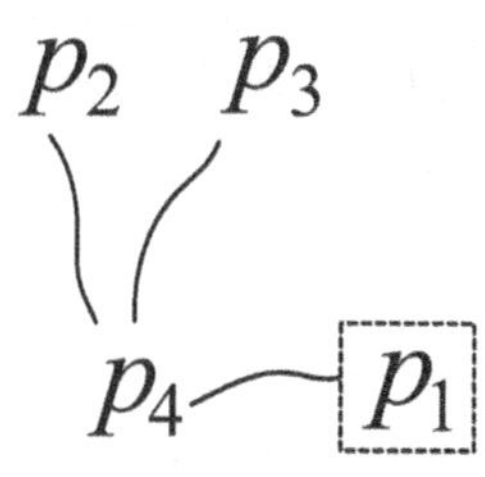

Fig. 7.6 The correlation network of the factors in Example 7.3

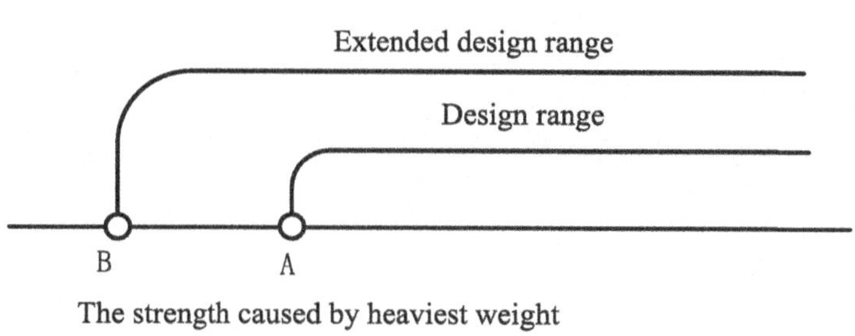

Fig. 7.7 Extension of design range

be affected by the configuration of the chassis (DP_1), rendering the coupled design decoupled. Consequently, the design matrix is updated to one that is diagonal one, thus resolving the coupling problem.

7.6 Discussion and Summary

This chapter took inspiration from the relevant AD theorems that are correlative to decoupling and identified 3 directions for resolving coupled problems. After the design function was built and the coupling type determined, Extenics was applied to establish the basic-element model that described the coupled problem. The basic-element models are formulated using the ordered triple expressions and the general set of characteristics in the basic-element are obtained with the semantic of logic. The uniform expressions provide thinking orientations for designers to analyze the coupling problems. The decoupling methodology has the descriptions as its foundation. The coupling expressed as a basic-element has the extension characteristics, based on which the decoupling methodology was systematized to transform the relationships for decoupling. With the decoupling directions of AD, three decoupled models each exploring AD and Extenics for following a particular decoupling direction were considered. Through the illustrative examples and corresponding analyses, it was shown that Extenics is applicable to describe coupling and the decoupling models were well-organized in expressing the coupled problems, thus helping solve the problem. The decoupling methodology system of Extenics provides designers with a large selection of methods and tools for transforming a coupled design problem to one that implies uncoupled or decoupled solutions.

References

Li W. J., Song Z. H., Mao E. R, & Suh S. (2018). Using Extenics to describe coupled solutions in axiomatic design. *Journal of Engineering Design, 30*, 1–3.

Suh, N. P. (2001). *Axiomatic Design: Advances and Applications*. Oxford University Press.

Chapter 8
Closure

8.1 Essence of the Synergized Theory

The goal of the principles of innovative design thinking is to enable designers and problem-solvers to be better thinkers. In this context a better thinker is one who is characteristically creative and innovative in generating multiple optimal solutions to difficult problems of high degree of complexity. The thinking process presented in the book is structured and well-defined using physical variables and mathematical mapping. Design thinking is a topic that includes need identification, problem framing-reframing, solution ideation, and creative thinking. It is an iterative process for identifying real need, questioning assumption, and redefining the problem. Design thinking as defined in the context of the theory is particularly powerful for addressing problems of ambiguity that are ill-defined and difficult to solve with standard methods. Its significance is in impacting the novelty, quality, implementation, value, and cost of the solution.

Need identification and analysis are the first step in design thinking. The step is critical as it allows the true need to be identified and bias toward finding solution to be avoided or ultimately eliminated. Abstraction, critical parameter identification, and questioning are the essential elements in this step.

The principles of design thinking are the result of exploring the Extenics and Axiomatic Design theories for features that are mutually enhancing. It is shown that the two theories are mutually complementary. The Axiomatic Design theory provides the framework for design element and the design process to be followed, whereas the theory of Extenics is particularly suitable for solving wicked problems. The basic-element models for functional requirement, design parameter, and coupling problem are well-defined with specific characteristics. The various extension approaches provided by Extenics are the manifestation of applying Chinese logic to thinking and reasoning—the conceptual thought process followed by Chinese philosophers and scholars. The meta thought gives rise to the basic concept of basic-element model and the transformation thought is embodied by the development of the extension methods and associated tools.

W. Li et al., *Principles of Innovative Design Thinking*,
https://doi.org/10.1007/978-981-19-0485-1_8

8.2 Implication and Outlook

Wicked problems are ill-defined, complex, and indeterminate, with the characteristics of being incomplete, shifting, contradicting and interdependent information that is difficult to gather. They are societal in nature, meaning that there are no straightforward solutions or clear answers to them. Wicked problems are ubiquitous. They are found in not just education but also finance, security, climate change, and monetary stability, to name only a few. Tackling wicked problems carries paramount implications to easing geopolitical tension, mitigating trade dispute, and reducing military conflict—the pressing issues that are threatening the order and stability of the world as we know it. The design thinking methodology and the associated design principles introduced in the book facilitate the mode of thinking that enables policy makers and solution providers to generate multitudes of equally viable solutions to wicked problems.

The synergy of Extenics and Axiomatic Design has the features of reframed decomposition, basic-element, and transformation method. Moreover, the reframed design process is well-structured with specific steps, rendering it feasible to be codified and implemented computationally using digital computers. Elements in the design process such as functional requirement, design parameter, design coupling, and design constraints are redefined using the specific characteristics. Well-structured and clearly defined, these elements are the foundation for storing knowledge and for generating innovative knowledge base—the two primary qualities characteristic of all the intelligent design systems and invention machines that are either available or currently under development. In other words, the theory of design thinking as presented in the volume carries the potential to be of significant while also far-reaching implications to artificial intelligence (AI) design and its broad applications.

Index

W. Li et al., *Principles of Innovative Design Thinking*,
https://doi.org/10.1007/978-981-19-0485-1

The manufacturer's authorised representative in the EU is Springer Nature Customer Service Centre GmbH, Europaplatz 3, 69115 Heidelberg, Germany. If you have any concerns regarding our products, please contact ProductSafety@springernature.com

Printed and bound by CPI Group (UK) Ltd, Croydon, CR0 4YY
30/09/2026
02225259-0015